New Studies of Polymer Surfaces and Interfaces

New Studies of Polymer Surfaces and Interfaces

Guest Editors

Xin Chen
Japan Trivedi
Zheyu Liu
Jianbin Liu
Zhe Li

Basel • Beijing • Wuhan • Barcelona • Belgrade • Novi Sad • Cluj • Manchester

Guest Editors

Xin Chen
College of Petroleum
Engineering
Xi'an Shiyou University
Xi'an
China

Japan Trivedi
Department of Civil and
Environmental Engineering
University of Alberta
Edmonton
Canada

Zheyu Liu
College of Petroleum
Engineering
China University of
Petroleum Beijing
Beijing
China

Jianbin Liu
College of Petroleum
Engineering
Xi'an Shiyou University
Xi'an
China

Zhe Li
School of Chemical
Engineering & Technology
China University of Mining
and Technology
Xuzhou
China

Editorial Office
MDPI AG
Grosspeteranlage 5
4052 Basel, Switzerland

This is a reprint of the Special Issue, published open access by the journal *Polymers* (ISSN 2073-4360), freely accessible at: www.mdpi.com/journal/polymers/special_issues/UO0VH25B80.

For citation purposes, cite each article independently as indicated on the article page online and using the guide below:

Lastname, A.A.; Lastname, B.B. Article Title. *Journal Name* **Year**, *Volume Number*, Page Range.

ISBN 978-3-7258-3956-8 (Hbk)
ISBN 978-3-7258-3955-1 (PDF)
https://doi.org/10.3390/books978-3-7258-3955-1

Contents

About the Editors

Xin Chen

Dr. Xin Chen, a lecturer at Xi'an Shiyou University, is mainly engaged in the research of multiphase fluid percolate theory in porous media, chemical and gas flooding for enhancing oil recovery theory, and unconventional oil and gas reservoirs for efficient development methods. Dr. Chen hosts one National Natural Science Foundation of China Project (Youngh), four provincial and ministerial projects such as the National Postdoctoral Research Program, and two school-enterprise cooperation projects. He also participated in the completion of three sub-projects of the major national science and technology projects of China and one National Natural Science Foundation of China (general project). He has published more than 40 SCI/EI academic papers, including 9 SCI papers as the first author (4 in the top journals of the first zone of the Chinese Academy of Sciences, 3 in the top journals of the second zone of the Chinese Academy of Sciences, and 1 ESI highly cited paper, with a total impact factor of 53.439), 5 EI papers, and 2 Scopus papers. Dr. Chen has been granted nine China invention patents and one US invention patent; he participated in the compilation of one standard of the oil and gas industry of the People's Republic of China; and he served as a reviewer for top academic journals in the industry at home and abroad, such as *Fuel, JPSE, Petroleum Science,* and *Physics of Fluids.*

Japan Trivedi

Dr. Japan Trivedi is a professor at the Faculty of Engineering, Civil and Environmental Engineering Dept, University of Alberta. His research interests include polymer (ASP) characterization, chemical EOR, functional polymer, rheology, and polymer-enhanced foams.

Zheyu Liu

Dr. Zheyu Liu is an associate professor at the College of Petroleum Engineering, China University of Petroleum, Beijing. Their research interests include enhanced oil recovery during chemical flooding, emulsification behavior characterization and control, multiphase fluid microscale flow, etc.

Jianbin Liu

Dr. Jianbin Liu is a lecturer at the College of Petroleum Engineering, Xi'an Shiyou University. Their research interests include injection/production profile modification, surfactant chemistry, enhanced oil recovery, polymer gels, and interfacial phenomena.

Zhe Li

Dr. Zhe Li is an associate professor at the School of Chemical Engineering and Technology, China University of Mining and Technology. Their research interests include enhanced oil recovery, conformance control, amphiphilic polymers, emulsion, and polymer gels.

Preface

Polymers are widely used in various industries. They can be used as a great oil recovery (EOR) system during chemical flooding. Polymer flooding, polymer/surfactant flooding, and polymer/surfactant/alkali flooding based on polymers are widely used worldwide and are of great significance for improving oil recovery in old oil fields (high water cut). Although the application of polymers is a relatively known method, the mechanisms of EOR from the perspective of surfaces/interfaces of polymers in reservoirs still requires further research and understanding.

Therefore, this reprint focuses on the surface/interface changes and physical and chemical reactions during the application of polymers. It reveals the mechanism of the surface and interface during the application of polymers, provides a theoretical basis and development ideas for the design and development of polymers, and promotes the rapid development of environmentally friendly, high-performance, and low-cost polymer applications.

Xin Chen, Japan Trivedi, Zheyu Liu, Jianbin Liu, and Zhe Li
Guest Editors

Article

Investigation of Polymer-Assisted CO_2 Flooding to Enhance Oil Recovery in Low-Permeability Reservoirs

Xin Chen [1,2,3,*], Yiqiang Li [4,*], Xiaoguang Sun [5], Zheyu Liu [4], Jianbin Liu [1,2,3] and Shun Liu [1,2,3,*]

1   College of Petroleum Engineering, Xi'an Shiyou University, Xi'an 710065, China; jianbin@xsyu.edu.cn
2   Shaanxi Key Laboratory of Advanced Stimulation Technology for Oil & Gas Reservoirs, Xi'an 710065, China
3   Ministry of Education Engineering Research Center of Development and Management for Low to Ultra-Low Permeability Oil & Gas Reservoirs in West China, Xi'an 710065, China
4   State Key Laboratory of Oil and Gas Resources and Exploration and College of Petroleum Engineering, China University of Petroleum (Beijing), Beijing 102249, China; zheyu.liu@cup.edu.cn
5   PetroChina Coalbed Methane Company Limited, Beijing 100028, China
*   Correspondence: cz_cup0709@163.com (X.C.); yiqiangli@cup.edu.cn (Y.L.); liushun631@163.com (S.L.)

Abstract: CO_2 flooding is a favorable technical means for the efficient development of low-permeability reservoirs, and it can also contribute to the realization of net-zero CO_2 emissions. However, due to the unfavorable viscosity ratio and gravity overriding effect, CO_2 channeling will inevitably occur, seriously affecting its storage and displacement effects. This paper conducts a systematic study on the application of polymer-assisted CO_2 flooding in low-permeability reservoirs. Firstly, the polymer agent suitable for low-permeability reservoirs is optimized through the viscosity-increasing, rheological, and temperature- and salt-resistant properties of the solution. Then, the injectivity performance, resistance-increasing ability, and profile-improving effect of the polymer solution were evaluated through core experiments, and the optimum concentration was optimized. Finally, the enhanced oil recovery (EOR) effects of polymer-assisted and water-assisted CO_2 flooding were compared. The results show that the temperature-resistant polymer surfactant (TRPS) has a certain viscosity-increasing performance, good temperature resistance performance, and can react with CO_2 to increase the solution viscosity significantly. Meanwhile, TRPS has good injection performance and resistance-increasing effect. The resistance increasing factor (η and η') of TRPS-assisted CO_2 flooding increases with increased permeability, the concentration of TRPS solution, and injection rounds. Considering η' and the profile improvement effect comprehensively, the application concentration of TRPS should be 1000 mg/L. The EOR effect of TRPS-assisted CO_2 flooding is 8.21% higher than that of water-assisted CO_2 flooding. The main effective period is in the first and second rounds, and the best injection round is three. The research content of this paper can provide data support for the field application of polymer-assisted CO_2 flooding in low-permeability reservoirs.

Keywords: polymer-assisted CO_2 flooding; CO_2 responsive; injectivity; resistance increasing; profile control; EOR

Citation: Chen, X.; Li, Y.; Sun, X.; Liu, Z.; Liu, J.; Liu, S. Investigation of Polymer-Assisted CO_2 Flooding to Enhance Oil Recovery in Low-Permeability Reservoirs. *Polymers* **2023**, *15*, 3886. https://doi.org/10.3390/polym15193886

Academic Editor: Leonard Ionut Atanase

Received: 7 August 2023
Revised: 9 September 2023
Accepted: 18 September 2023
Published: 26 September 2023

1. Introduction

With increased international energy consumption and oil exploration and development intensification, low-permeability reservoirs have contributed to Chinese oil and gas resource production. Developing low-permeability oilfields is generally dominated by water injection, and gas flooding, especially CO_2 flooding, is an efficient development technology for low-permeability reservoirs with excellent prospects [1,2]. CO_2 flooding can play the role of high-efficiency oil displacement and geological storage simultaneously, in line with the policy of carbon peak and carbon neutrality, and has good application advantages. However, due to unfavorable mobility ratio and gravity overriding, CO_2 will channel in the reservoir, significantly reducing the development effect [3–5]. Therefore, an economical and effective method for mitigating gas channeling is the key to further improving CO_2 flooding in low-permeability reservoirs. Polymer flooding is the primary EOR technology in China [6], and it has been successfully applied in Daqing Oilfield, Xinjiang Oilfield, Dagang Oilfield, etc. [7,8]. Polymers can reduce the fluidity of injected fluids, reduce the permeability of advantageous channels through adsorption and retention, and ultimately expand the swept volume [9,10]. The resistance coefficient and residual resistance coefficient are usually used to evaluate the injectivity performance of polymer solutions, and having a suitable resistance coefficient and residual resistance coefficient is the prerequisite for the successful application of polymers [11]. Indoor research results show that polyacrylamide (HPAM) polymers can increase oil recovery by 10–12% of OOIP after water flooding [12]. Meanwhile, the molecular modification of the polymer can play an interface effect, stripping, dispersing, and carrying oil droplets [13]. In addition, polymer gels [14], polymer microspheres [15–17], polymeric surfactants [18], and other agents [19] can play a role in profile control and flooding, further improving oil recovery. A comparison of the EOR effects of different enhanced CO_2 injection methods is shown in Appendix A Table A1. It can be found that as the system strength increases, the EOR gradually increases. However, due to the high viscosity characteristics of polymers, it must keep a good reservoir matching relationship while exerting the viscosity-increasing effect [20,21], so it is usually used in medium-high permeability reservoirs.

It is of excellent research significance to explore using polymers to assist CO_2 flooding in low-permeability reservoirs to improve gas channeling and expand the swept volume. Dann et al. [22] compared the injection performance of polymer solutions in different low-permeability cores through core flooding experiments and pointed out that reservoir core characteristics other than permeability can dominate polymer performance. In addition, Ghosh et al. [23] conducted polymer injectivity experiments on two cores with a permeability of 15 mD but with significant differences in pore throat distribution. The results showed that the polymer could flow smoothly in the cores with a bimodal distribution of pore throats. But in another core with unimodal distribution, the flow is difficult. The above two studies show that the polymer solution has the potential to flow smoothly in the low-permeability reservoir, but it is necessary to meet the matching of the polymer and the reservoir. Marliere et al. [24] used cores with an effective permeability of 1–3 mD to carry out oil displacement experiments with polymers (molecular weight of 3 million Da and viscosity of 2 cP), which can increase oil recovery by 20% to 40% based on water flooding. Bennetzen et al. [25] proved by experiments that partially hydrolyzed polyacrylamide (HPAM, with a molecular weight of 8 million Da and a concentration of less than 5000 mg/L) solution injected into a 0.3 mD carbonate reservoir core at a rate of 0.5 mL/h will not produce core plugging. Mohammed Taha et al. [26] studied the application of low-salinity polymer flooding in high-temperature, high-salt, and low-permeability reservoirs. The polymers can maintain a viscosity of 2–3 cP (2500–4000 mg/L), with good injection and enhanced oil recovery effects. Leon et al. [27] reported a successful field pilot of the polymer in the low-permeability Palogrande-Cehu field. Considering the low permeability of the reservoir (6–150 mD), three polymers with different molecular weights (average concentration 1021 mg/L, viscosity 3.43 cP) were used, and the viscosity change was monitored. It is estimated that EOR efficiency can reach up to 8%, and the water cut of some wells is reduced by as much as 14%. The above studies show that the successful application of polymers in low-permeability reservoirs requires low viscosity and low injection velocity. For CO_2 flooding in low-permeability reservoirs, a CO_2 response mechanism based on the above studies [28–30] can be used to prevent gas channeling. Inject a low-viscosity polymer surfactant into the reservoir, which can react with CO_2 to form worm-like micelles, significantly increasing the viscosity of the system [31,32]. However, due to the high viscosity of this type of micelles, it is mainly used for the plugging of fractures in low-permeability reservoirs [33–35]. Through the above literature review, two problems exist in the current research on polymer-assisted CO_2 flooding in low-permeability reservoirs: (1) how to optimize the polymer agents with reservoir adaptability; and (2) how to apply the CO_2 response characteristics in low-permeability reservoirs matrix.

Aiming at the above problems, the present paper studies a temperature-resistant CO_2-responsive polymer surfactant (TRPS) to assist CO_2 flooding to improve the oil recovery of low-permeability reservoirs. The advantages of it as an assisted agent were evaluated by comparing its static properties, such as viscosifying properties, rheological properties, and temperature and CO_2 resistance with xanthan gum. Then, the injection performance, resistance-increasing performance, and profile improvement effect of TRPS were evaluated through core injectability experiments, and the optimal injection concentration was determined. Finally, the enhanced oil recovery effects of water-assisted CO_2 flooding and TRPS-assisted CO_2 flooding were compared. The research in this paper can provide the experimental basis and data support for agent selection, performance evaluation, and injection parameter optimization in the application process of polymer-assisted CO_2 EOR of low-permeability reservoirs.

2. Material and Method

2.1. Material

Polymers: The injectivity performance of polymer solution in the low-permeability reservoirs can only be satisfied when the polymer molecular weight and solution viscosity are low enough. Low-molecular-weight polymers and polymeric surfactants can be used as two typical low-molecular and low-viscosity polymer systems. Here, xanthan gum (molecular weight is about 5×10^6 Da) and a kind of CO_2-responsive temperature-resistant polymeric surfactant (molecular weight is about 10^6 Da) are selected as representatives of two polymer systems to conduct the following research. TRPS with certain interfacial activity and viscosity-increasing properties was obtained by grafting functional functional groups onto the main chain of polyacrylamide. Meanwhile, it has the CO_2 corresponding characteristics and can significantly increase the viscosity of the solution in a CO_2 environment.

Liquids: The inorganic salt purchased from Shanghai Aladdin Reagent Co. (Shanghai, China) was proportionally added to the deionized water (DI water) to prepare the simulated formation water, and the total salinity was 7000 mg/L. DI water was prepared by UPT-I-10T Ultra-pure Water Purifier from Chengdu Youpu Super Pure Technology Co. (Chengdu, China). The simulated oil is white oil purchased from, Shanghai Aladdin Co. (Shanghai, China), with a viscosity of 15.5 cP at 80 °C (target reservoir temperature).

Cores: Cylindrical Berea cores with a size of 3.8×30 cm were used for polymer-assisted CO_2 flooding experiments, and the gas permeabilities are about 5 mD, 10 mD, and 20 mD, respectively. The effective permeability of each core was specifically tested before each experiment.

2.2. Polymers Static Properties

2.2.1. Viscosity-Increasing Properties

The viscosity-increasing property of a polymer refers to the ability of a polymer to dissolve in water to increase the viscosity of the aqueous phase and is the most critical parameter for evaluating a polymer agent. Use DI water to prepare XG and TRPS mother liquor with a concentration of 5000 mg/L, and stir for 2 h with an electronic stirrer at 400 rpm. Then, the mother liquid could be diluted using simulated formation water to prepare the target polymer solution (100 mg/L, 300 mg/L, 500 mg/L, 700 mg/L, and 900 mg/L). The viscosities of the polymer solutions will be tested by a Brookfield viscometer after stirring for 1 h with an electronic stirrer at 200 rpm. The shear rate of the Brookfield viscometer is 7.4 1/s, and the test temperature is 80 °C.

Next, 100 mL of the XG and TRPS solutions with the target concentration were poured into the Warring agitator, respectively, sheared at 3500 rpm for 1 min, and then we tested the viscosity of the solution again and calculated the viscosity retention rate.

2.2.2. Polymer Rheological Properties

The viscosity of polymer solution decreases with the increase in shear rate, which has a typical shear thinning property. Evaluating the viscosity of polymers at different shear rates is of great significance for testing their EOR efficiency. The rheological curves of XG and TRPS solutions at concentrations of 100 mg/L, 300 mg/L, 500 mg/L, and 1000 mg/L were tested using a Haake rheometer (HAAKE RS6000, Thermo Electron Karlsruhe GmbH, Karlsruhe, Germany), and the shear rate was 0.1 1/s to 1000 1/s, and the test temperature was 80 °C.

2.2.3. Polymer Reservoir Adaptability

After the polymer is injected into the reservoir, it needs to stay in the reservoir environment for a long time, and its performance under this condition determines its effect. The temperature resistance and CO_2 resistance performance requirements of the polymer solution are evaluated by the viscosity retention rate to meet the application conditions of the oil reservoir. Put XG and TRPS solution with a concentration of 200 mg/L, 500 mg/L, 800 mg/L, 1000 mg/L, and 1500 mg/L in a high-temperature aging tank separately and aged in an oven at 80 °C for 30 days. Samples were taken at regular intervals (2, 4, 6, 8, 10, 15, 20, and 30 d) to test the viscosities using a Brookfield viscometer. In addition, injecting CO_2 into piston containers with polymer solutions of 300 mg/L, 500 mg/L, and 800 mg/L until the pressure reaches 10 MPa. Then, seal the piston and place it at 80 °C for ten days for a certain period to test the viscosity of the solution.

2.3. The Flow Performance of TRPS in Low-Permeability Reservoirs

2.3.1. The Injectivity Ability of TRPS

The injectivity performance of polymer solution is mainly quantitatively evaluated by Resistance Factor (R_F) and Residual Resistance Factor (R_{FF}). The specific experimental procedure is as follows: (1) After the core was vacuumed for 3 h, it was saturated with simulated water by self-priming for more than 4 h, and the porosity was calculated; (2) According to Figure 1a, the simulated formation water was injected with the ISCO pump at a constant rate, and the pressure difference between the two ends of the core was recorded as ΔP_1 after the pressure became stable, and the core water permeability was calculated using Darcy law; (3) The ISCO pump was used to inject TRPS solution at a constant rate until the pressure is stable, and the pressure difference of the core was recorded as ΔP_2; and (4) The ISCO pump was used to inject simulated formation water at a constant rate until the pressure was stable, and the pressure difference at both ends of the core was recorded as ΔP_3. The injection rate is 0.3 mL/min, and the TRPS solution needs to be injected continuously for at least 2 PV. Finally, the FR = $\Delta P_2 / \Delta P_1$ and the FRR = $\Delta P_3 / \Delta P_1$ can be calculated. The specific experimental scheme is shown in Table 1.

In addition, a zeta potential analyzer (Zetasizer Nano ZS, Malvern Panalytical, malvern city, England) was used to test the hydrodynamic size of TRPS solutions with different concentrations, and the Poiseuille formula was used to calculate the mean pore throat size of the cores. The injectivity performance results of TRPS can be analyzed by comparing the above two sizes.

Table 1. The experimental scheme of TRPS injectivity properties.

Polymer Concentration, mg/L	Gas Permeability, mD	Water Permeability, mD	Core Volume, cm^3	Pore Volume, cm^3	Porosity, %
1000	5	2.99	226.7	24.32	10.73
500	10	5.36	226.7	34.28	15.12
1000	10	5.23	226.7	35.82	15.8
2000	10	5.38	226.71	33.56	14.80
1000	20	10.12	226.71	43.80	19.32

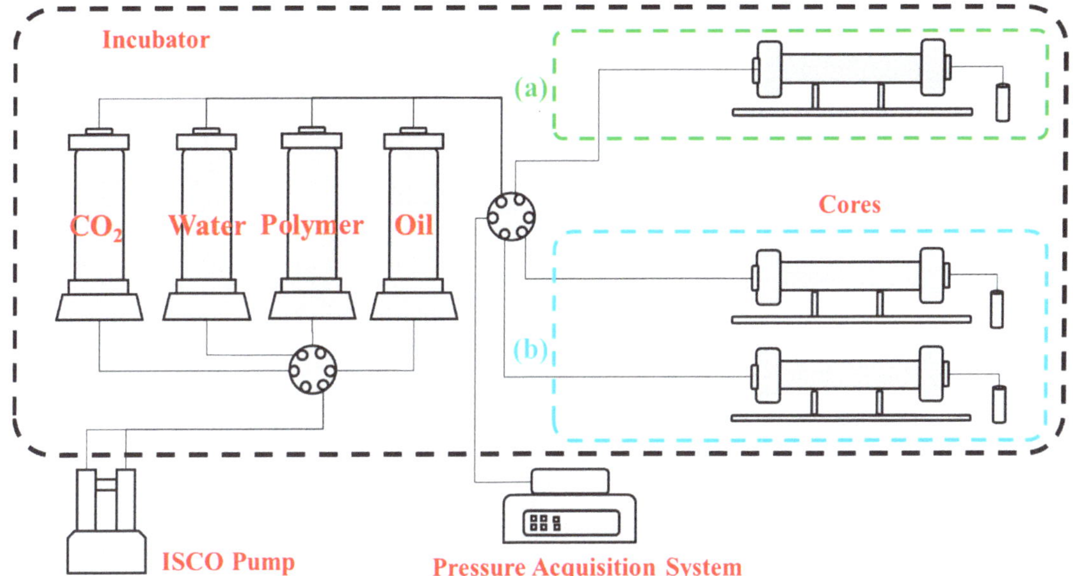

Figure 1. The flowchart of the flow performance of the TRPS solution. (**a**) Single core flooding flow chart, suitable for injectivity ability testing, resistance-increasing performance testing, and EOR efficiency evaluation. (**b**) Two-core parallel core flooding flow chart, suitable for profile control performance evaluation.

2.3.2. Resistance-Increasing Performance of TRPS-Assisted CO_2 Flooding

After clarifying the injectivity performance of TRPS, it is necessary to evaluate the resistance-increasing effect of TRPS in the presence of CO_2. Evaluation is also carried out by RF and RFF. The specific experimental procedure is the same as that in Section 2.3.1, except that step (3) is to inject CO_2 and TRPS solution into the core simultaneously using the ISCO pump. The specific experimental scheme is shown in Table 2.

Table 2. Scheme of the evaluation of the resistance-increasing performance of TRPS.

Solution	Viscosity, cP	Injection CO_2	Gas Permeability, mD	Water Permeability, mD	Porosity, %
TRPS	9.3	Yes	10	5.20	15.51
HPAM	10.1	No	10	5.36	14.98
TRPS	9.3	No	10	5.32	15.11

Meanwhile, it is necessary to evaluate the resistance-increasing performance of the TRPS-assisted CO_2 flooding process. Berea cores with a size of 3.8 cm × 20 cm and a permeability of 5 mD, 10 mD, and 20 mD were selected to evaluate the resistance-increasing effect of TRPS. The specific experimental process is as follows: (1) The core was vacuumed for 3 h and then saturated with water for more than 4 h by self-priming and calculating the porosity; (2) Connected the experiment flow chart in Figure 1a and carried out water flooding and gas flooding at a constant rate using the ISCO pump until the pressure is stable. Recorded the stable pressure P_0 as the basic gas flooding pressure; (3) Injected TRPS solutions with concentrations of 200 mg/L, 500 mg/L, and 1000 mg/L at a constant rate, and recorded the stable pressure P_1, respectively; (4) Carried out gas flooding again at a constant injection rate to record the pressure P_1' before gas channeling; and (5) Repeated Step (3) and Step (4) multiple times to obtain the pressures P_i and P_i'. TRPS resistance increase coefficient $\eta_{(i)} = P_i/P_0$ and resistance increase coefficient of gas flooding $\eta'_{(i)} = P'_i/P_0$.

2.3.3. Profile Control Performance of TRPS

The 3.8 cm $\times$ 20 cm Berea cores with permeability of 5 mD and 20 mD were connected in parallel to simulate reservoir heterogeneity (permeability difference ratio is 4) to carry out the TRPS-assisted CO_2 flooding experiment without oil. The specific experimental procedure is as follows: (1) After the core is evacuated for 3 h, it is saturated with simulated water by self-priming for more than 4 h, and the porosity is calculated; (2) The simulated formation water was injected at a constant rate by an ISCO pump, and the pressure at both ends of the core is recorded after the pressure is stable. The stable pressure difference is recorded as ΔP_1, and the core water permeability is calculated by Darcy law; (3) According to Figure 1b, the experimental process was connected and simulated water and CO_2 were co-injected using the ISCO pump. Recorded the liquid production conditions and injection pressure of the cores; (4) TRPS and CO_2 were co-injected using the ISCO pump at a constant rate until the pressure is stable; and recorded the liquid production conditions and injection pressures of the two cores. The combined injection rate of liquid and gas is 0.6 mL/min (1:1 volume ratio), and if there is no CO_2 injection, the TRPS injection rate is 0.6 mL/min. The specific experimental scheme is shown in Table 3.

Table 3. TRPS profile control effect experimental scheme.

Solution	Concentration, mg/L	Injection CO_2	Gas Permeability, mD	Water Permeability, mD
TRPS	500	Yes	5/20	3.12/9.98
TRPS	1000	No	5/20	3.03/10.20
TRPS	1000	Yes	5/20	3.21/10.15

2.4. EOR Efficiency of TRPS-Assisted CO_2 Flooding

The oil displacement experiments were carried out using Berea cores (gas permeability is 10 mD and the water permeability is 5.19 mD). Comparing the oil recovery and pressure change law of water-assisted CO_2 flooding and TRPS-assisted CO_2 flooding, the advantages of TRPS-assisted CO_2 flooding in improving oil recovery were clarified, and the best effect period was determined. The specific experimental procedure is as follows: (1) After the core is evacuated for 3 h, it is saturated with water by self-priming for more than 4 h, and the porosity is calculated; (2) The simulated formation water is injected at a constant rate by an ISCO pump, and the pressure at both ends of the core is recorded after the pressure is stable. The pressure difference is ΔP_1, and the core water permeability is calculated by Darcy law; (3) According to the connection experiment process in Figure 1a, the ISCO pump is used to inject crude oil into the core at a constant rate until there is no water production, and the oil saturation is calculated; (4) Simulated formation water was injected with ISCO pump at a constant rate until the water cut reaches 90% and then switched to CO_2 flooding until gas channeling without producing oil; (5) Injected TRPS solution (1000 mg/L) with ISCO pump at a constant rate of 0.2 PV and then switched to CO_2 flooding until no oil is made again, this is a round of TRPS-assisted CO_2 flooding process; (6) Repeated step (5) for a total of three times, that is, three rounds of TRPS-assisted CO_2 flooding are completed. During the experiment, the fluid production and injection pressure of the core were recorded. The injection rate of liquid and gas was 0.3 mL/min. During the experiment, a back pressure of 10 MPa was applied at the outlet end of the core by a back pressure valve.

3. Result and Discussion

3.1. Static Performance Comparison between XG and TRPS

3.1.1. Viscosity-Increasing Performance

Figure 2 shows the relationship curves of the viscosity of XG and TRPS polymer solutions with the change of solution concentration at 80 °C. The results show that both polymers have certain viscosity-increasing properties in terms of demand in the field of oil and gas field development engineering, and the viscosity-increasing effect of XG is more significant. When the solution concentration exceeds 300 mg/L, the viscosity-increasing effect of the XG solution is significantly improved, much higher than that of TRPS at the same concentration. When the solution concentration was less than 1000 mg/L, the viscosity-increasing effect of TRPS gradually increased with the increase in the solution concentration, and the change was stable. When the concentration is 1000 mg/L, the viscosities of XG and TRPS solutions are 15.21 cP and 4.57 cP, respectively. Within this concentration range, XG and TRPS have the potential of low-permeability reservoir injectability [36], meeting the requirements of follow-up research. In addition, the viscosity of TRPS solutions with concentrations of 700 mg/L and 1000 mg/L is similar to that of XG solutions with concentrations of 100 mg/L and 300 mg/L, respectively. The performance of the two solutions at the same viscosity will be compared through the above two concentrations.

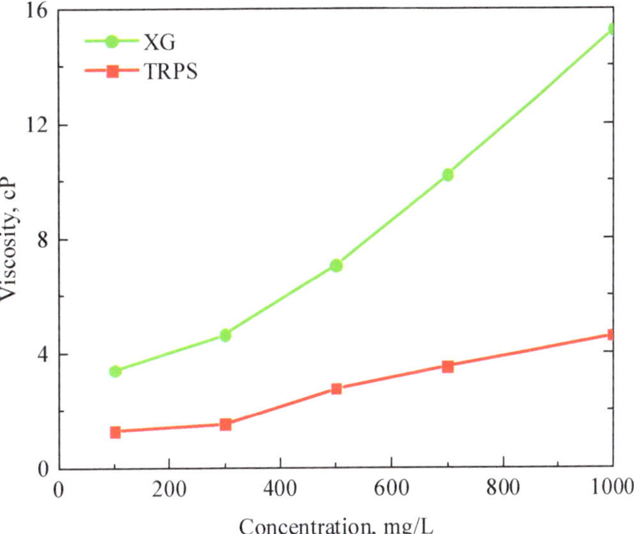

Figure 2. The viscosity-increasing ability of XG and TRPS.

Figure 3a,b show that the viscosities and retention rates of XG and TRPS solutions increase with the increase in solution concentration. The relationship between the viscosity retention rate and the solution concentration of XG and TRPS solutions after shearing at 3500 rpm is compared, as shown in Figure 3c. The viscosity retention rates of the two polymer solutions after shearing are both above 80%. Whether the viscosity is the same or the concentration is the same, the viscosity retention rate of TRPS is greater than that of XG because the viscosity-increasing mechanism of XG and TRPS mainly includes: (1) the polymer molecules in water are entangled to form a structure; (2) The hydrophilic groups in the polymer chains are solvated in water, and the apparent molecular volume of the polymer increases. XG has a more considerable molecular weight, and the larger molecular aggregates formed are more easily damaged by shearing.

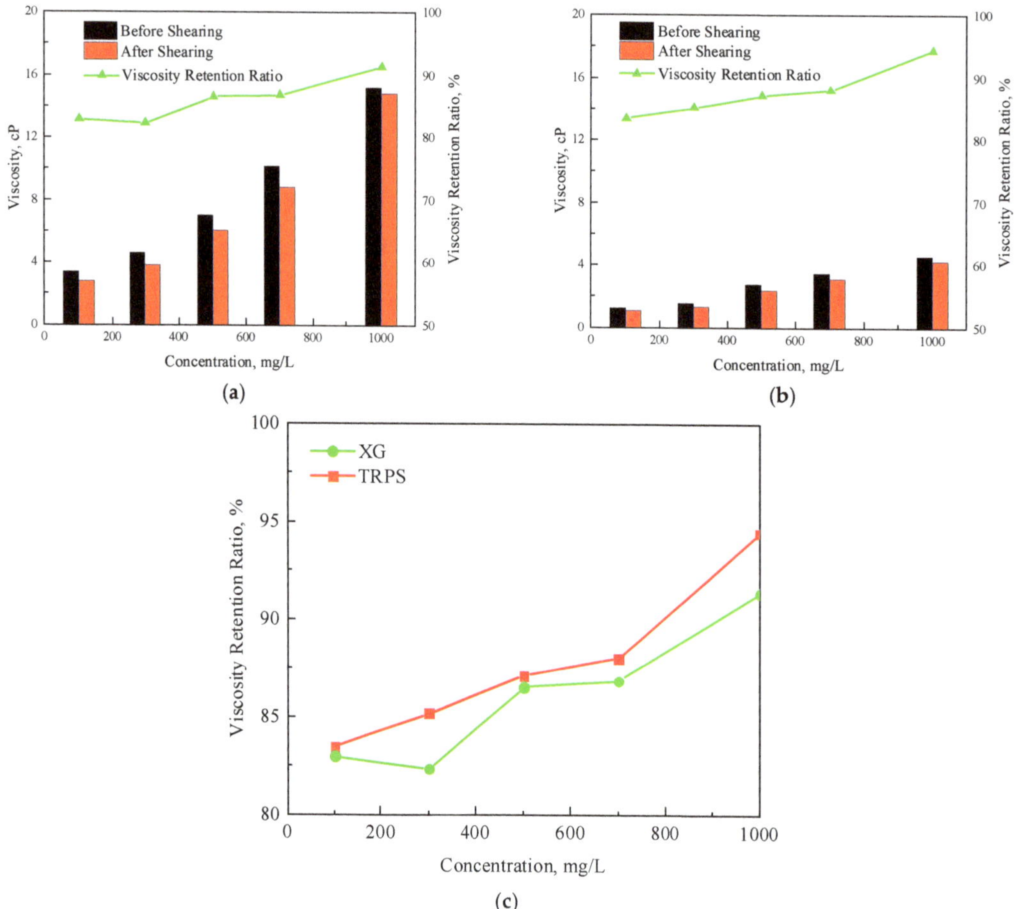

Figure 3. Viscosity and viscosity retention rate of polymer solution after shearing, (**a**) XG, (**b**) TRPS, (**c**) comparison of viscosity retention rate.

3.1.2. Rheological Properties

Figure 4 shows the rheological curves of XG and TRPS with four concentrations. Figure 4 shows that the rheological curves of XG and TRPS solutions are power-law curves, showing typical shear thinning characteristics. The shear rates corresponding to the rheological curves of XG and TRPS solutions reaching the second Newton zone are about 10 1/s and 5 1/s, respectively, and the molecular conformation transition has reached the limit value. Because in the practice of oil field production, the shear rate of the polymer system in the underground is about 1–10 1/s, combined with the rheological curve, it is easier for TRPS to reach the second Newton zone in this shear rate range. The adaptability to the reservoir shearing of TRPS is more robust. Meanwhile, comparing the rheological curves of TRPS (1000 mg/L) and XG (300 mg/L) solutions with the same viscosity, it can be found that the viscosity of XG solution is higher at low shear rate, and the viscosity of the two is similar in the second Newton zone. This is mainly because the molecular weight of XG is large, and the viscosity-increasing property of the molecular coil is stronger at a lower shear rate, and as the shear rate increases, the difference between the two molecular weights for the viscosity-increasing performance decreases.

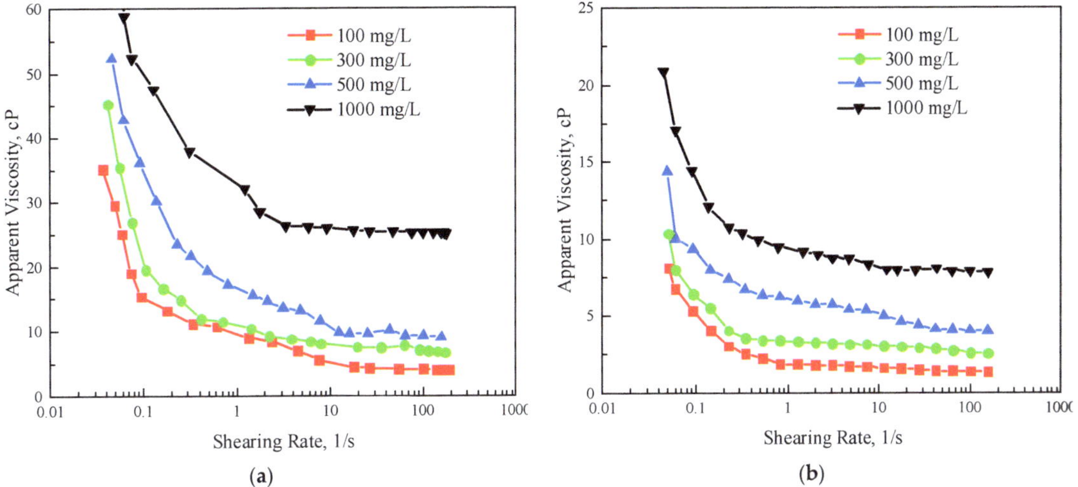

(a)

(b)

Figure 4. The rheological curve, (**a**) XG, (**b**) TRPS.

3.1.3. Temperature-Resistance and CO_2-Resistance Performance

Figure 5 compares the viscosity retention curves of XG and TRPS solutions after aging in high-temperature and CO_2 environments. Figure 5a shows that high temperature has a more significant effect on the viscosity of XG solution but has little impact on the viscosity of TRPS. The viscosity of the XG solution decreased rapidly with the increase in aging time, then reduced steadily after aging for ten days, and the viscosity retention rate after aging for 30 days was less than 20%. The viscosity of the TRPS solution decreased slowly with the increase in aging time, and the final viscosity retention rate was about 80%. This is because high-temperature conditions will gradually untangle the initially entangled polymer coils, and long-term aging will cause further breakage of the stretched molecular bonds, resulting in a significant decrease in the viscosity of the solution. However, TRPS has a temperature-resistant monomer and a short molecular coil, showing good temperature resistance.

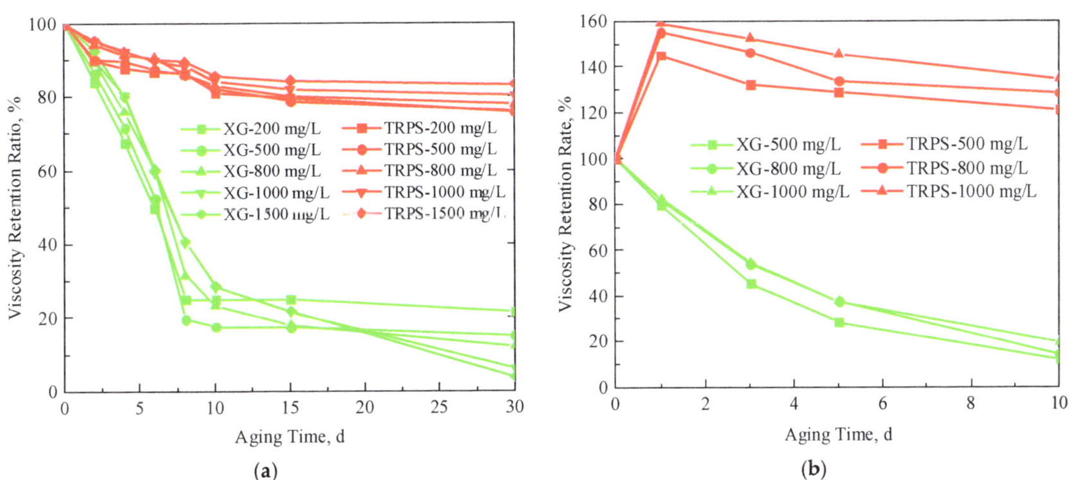

(a)

(b)

Figure 5. The viscosity retention ratio of XG and TRPS solutions, (**a**) high-temperature aging, (**b**) CO_2 environment.

Figure 5b shows that the presence of CO_2 significantly reduces the viscosity of the XG solution, and the viscosity retention rate drops to about 20% after aging for ten days. However, after aging in a CO_2 environment, the viscosity retention rate of TRPS solution increased first (up to more than 150%), then decreased slowly, and finally remained at about 120%. CO_2 will increase H^+ in the polymer solution and destroy the biomolecular chains in the XG solution, thereby losing the viscosity-increasing effect [37]. The TRPS molecular chain contains tertiary amine groups with CO_2 response characteristics. The primary tertiary amine group can undergo an acid-base neutralization reaction with dissolved CO_2 in aqueous solution to form a bicarbonate structure. The tertiary amine group is protonated to form a quaternary salt cation structure, and the molecular chain is hydrophilic enhanced. Therefore, the presence of CO_2 leads to larger TRPS molecular aggregates and an increase in the viscosity of the system [38,39]. The specific reaction mechanism is shown in Figure 6.

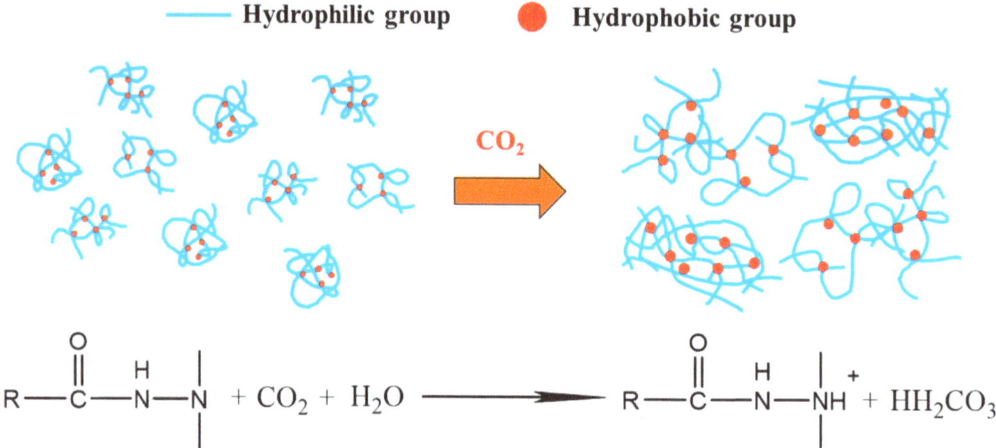

Figure 6. Schematic diagram of the mechanism of viscosity increase in response to TRPS and CO_2.

In addition, it can be seen from Figure 5 that regardless of the concentration of the TRPS solution, the viscosity retention rate after high-temperature aging and CO_2 environment aging is higher than that of XG. This also shows that under the same viscosity, the temperature resistance and CO_2 resistance of the TRPS solution are better than XG.

3.2. Injectivity of TRPS

Figure 7 shows the TRPS injection pressure curves under three concentrations and three core permeabilities. The injection pressure increased significantly during the TRPS flooding, and it decreased slightly and then remained stable in the subsequent water displacement stage. It also showed a certain resistance-increasing ability under the condition of no CO_2. R_F and R_{FF} increase with the increase in TRPS concentration and the decrease in core permeability. When the concentration of TRPS is fixed at 1000 mg/L, R_F and R_{FF} in 5 mD core are 1.36 and 1.28, respectively, indicating good injectivity. However, the R_F and R_{FF} of 1000 mg/L TRPS solution in 20 mD cores are both less than 1.1, which means that although its resistance-increasing effect is poor, its injectivity is good. R_F and R_{FF} of 500 mg/L, 1000 mg/L, and 1500 mg/L TRPS solutions in 10 mD cores are, respectively, distributed between 1.08–1.27 and 1.01–1.15, indicating that the concentration of TRPS solution will not significantly influence its injectivity. When the core permeability is greater than 5 mD, and the TRPS solution concentration is less than 1500 mg/L, R_{FF} is close to 1, indicating that the TRPS solution has less adsorption and retention in the formation, less damage to the formation, and has good injection performance.

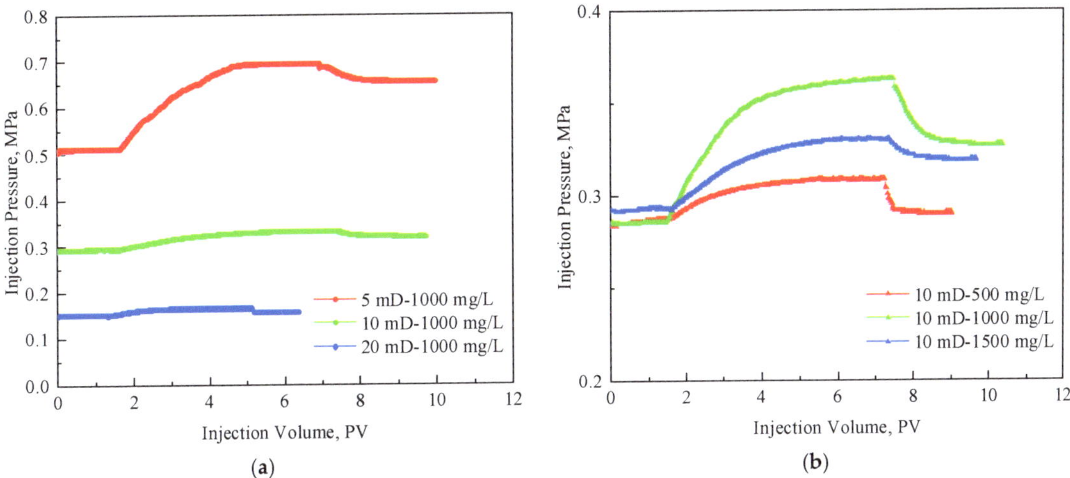

Figure 7. TRPS injection pressure curve. (**a**) The TRPS concentration is 1000 mg/L, and the core permeability is 5 mD, 10 mD, and 20 mD. (**b**) The core permeability is 10 mD, and the TRPS concentration is 500 mg/L, 1000 mg/L, and 1500 mg/L.

Figure 8 compares the hydrodynamic size of the TRPS solution and the mean pore throat size of the core. It can be found that as the concentration of TRPS solution increases, its hydrodynamic size gradually increases from 150 nm to 350 nm, and no obvious intermolecular association occurs. As the core permeability increases, the mean pore throat size increases from 470 nm to 650 nm. The mean pore throat size of TRPS is smaller than that of the core, ensuring its good injectivity performance. Meanwhile, TRPS molecules can effectively plug pore throats through the 1/2 and 1/3 bridging theory. This can explain why TRPS in Figure 7 can effectively increase the injection pressure while ensuring a low resistance coefficient.

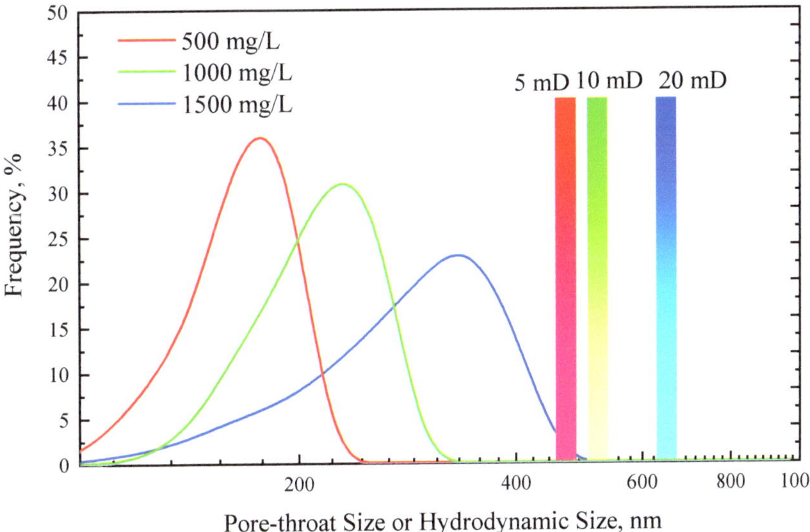

Figure 8. Comparison of the pore throat sizes of cores and the hydrodynamic sizes of TRPS solutions.

3.3. Resistance Increasing Ablity of TRPS

Figure 9 shows the injection pressure curves of the TRPS injection, HPAM injection, and co-injection of TRPS and CO_2. The injection pressure rises rapidly after HPAM injection, which is much higher than that of TRPS injection, showing that even if the viscosity of the polymer solution is the same, the molecular weight of the polymer also affects its injection performance and resistance-increasing performance. The R_F of HPAM injection is 7.67, which is not significant in value, but the injection pressure rises above 2.0 MPa, and there will be a problem with too high injection pressure during the oilfield application. Meanwhile, the R_{FF} of HPAM injection is 4.45, indicating that HPAM has a large amount of retention and adsorption in the low-permeability layer, which is not conducive to the subsequent development of the formation. However, the R_F and R_{FF} of TRPS injected are 1.13 and 1.09, respectively, and the resistance-increasing effect is poor. The injection pressure during the co-injection of TRPS and CO_2 increases significantly because the more complex molecular aggregates formed after the mixing of TRPS and CO_2 significantly increase its flow resistance (Figure 6). At this time, R_F and R_{FF} rose to 2.67 and 1.23, respectively, and the effect of increasing resistance was improved. Figure 9 shows that it is difficult for conventional polymers (relatively high molecular) to achieve the coordination of injectivity and resistance-increasing in low-permeability reservoirs, while the response of TRPS and CO_2 allows it to be injected into the reservoir smoothly and achieve resistance increase inside the reservoir effect.

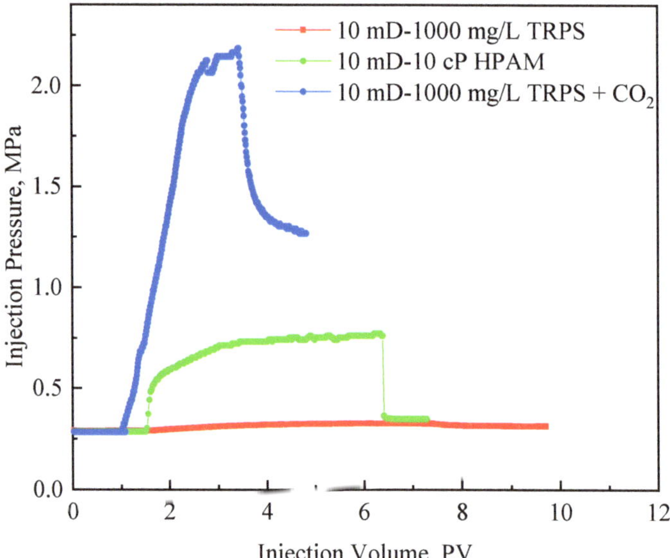

Figure 9. The injection pressure curves of TRPS injection, HPAM injection, and co-injection of TRPS and CO_2.

The injection pressures of multiple rounds of TRPS-assisted CO_2 flooding with three kinds of permeability and three kinds of concentrations are collected and the resistance increase coefficients η and η' of each round are shown in Table 4. Table 4 shows that η is greater than η' in the same round because the resistance of the injected liquid must be greater than the resistance of the injected CO_2. Under the same permeability conditions, η and η' increased significantly after increasing the concentration of TRPS. When the concentration is 1000 mg/L, η can reach more than three times that when the concentration is 200 mg/L. η' can reflect the resistance-increasing effect after the reaction of TRPS and

CO_2, so the relationship between η' of the above rounds and the change of core permeability is plotted into a resistance-increasing coefficient chart, as shown in Figure 10.

Table 4. η and η' of each round of TRPS-assisted CO_2 flooding.

Parameters			Round 1		Round 2		Round 3		Round 4	
Permeability, mD	Concentration, mg/L	Gas Channeling Pressure, MPa	η	η'	η	η'	η	η'	η	η'
	200	0.25	1.96	1.36	3.08	1.72	3.92	2.12	5.60	3.16
5	500	0.33	2.70	1.55	6.18	2.21	/	/	/	/
	1000	0.31	6.74	6.53	11.2	10.8	16.1	15.7	22.9	22.5
	200	0.13	2.20	1.73	3.23	2.20	4.96	2.60	9.37	4.17
10	500	0.15	5.00	2.60	9.53	3.47	/	/	/	/
	1000	0.14	8.43	6.93	12.4	7.86	20.0	10.0	28.0	13.2
	200	0.07	2.57	1.86	4.43	3.14	7.43	5.43	10.4	7.14
20	500	0.08	7.50	4.13	14.1	5.75	/	/	/	/
	1000	0.08	8.50	5.25	18.0	7.13	29.1	14.6	42.3	17.7

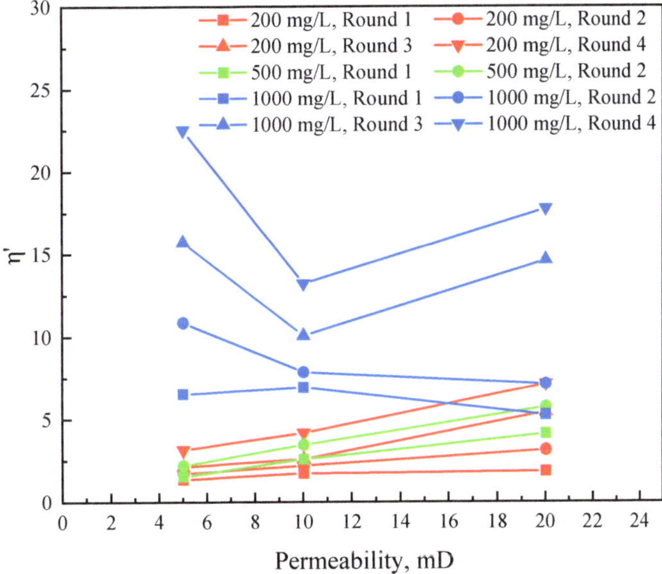

Figure 10. The resistance-increasing coefficient chart.

When the concentration of TRPS is lower than 1000 mg/L, the increase in core permeability and alternate rounds will increase η'. This is because although the increase in permeability will reduce the injection pressure of TRPS-assisted CO_2 flooding, the gas channeling pressure of CO_2 flooding will also decrease, eventually increasing η'. The molecular aggregates produced by the reaction of TRPS and CO_2 will be adsorbed and retained in the core, so η' increases with the increase in injection rounds. When the concentration of TRPS is 1000 mg/L, the relationship curves of η' change abnormally. This also shows once again the matching between TRPS and the reservoir (the low-permeability reservoir will not have unusually high pressure caused by the poor injectivity of molecular aggregates produced after the reaction of TRPS and CO_2) and the cumulative effect of multiple rounds of resistance-increasing performance. When the permeability is 5 mD, the molecular aggregates formed by the reaction of TRPS and CO_2 are poorly compatible with the core, resulting in an abnormal increase in injection pressure and a significant increase in η'. Then, the pore throats of the 10 mD and 20 mD cores became more prominent, and the flow capacity increased, which relieved the contradiction between the poor matching between TRPS and the core, and the change law of η' was normal. In addition, the irregularity of η' of the TRPS solution with a concentration of 1000 mg/L increases with the increase in injection rounds, mainly in the core with a permeability of 5 mD. This is due to the increase in retention of TRPS in the core due to the decrease in permeability.

3.4. Profile Control Effect of TRPS

Figure 11 shows the injection pressure and fractional flow rate curves of the TRPS-assisted CO_2 flooding in a two-core parallel model. Figure 11 shows that the injection pressure of 1000 mg/L TRPS-assisted CO_2 flooding is higher than that of TRPS flooding separately, which also leads to a more obvious reduction in the fractional flow rate of the increased permeability layer. The injection pressure of TRPS-assisted CO_2 flooding with a concentration of 500 mg/L is lower than that of TRPS separately (1000 mg/L), showing that TRPS needs to reach a certain concentration before fully reacting with CO_2. Combined with the resistance-increasing effect of TRPS, it can be determined that the optimal concentration of TRPS should be 1000 mg/L for reservoirs with a permeability above 5 mD and 500 mg/L for reservoirs below 5 mD.

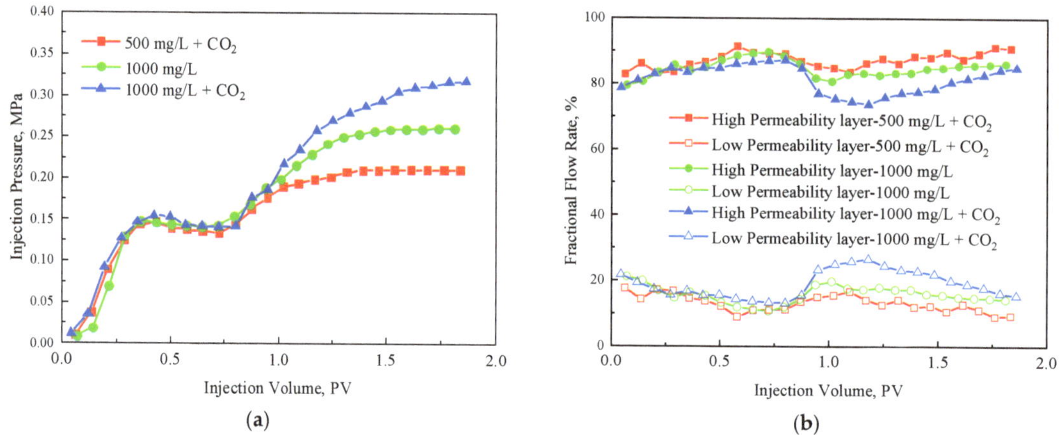

Figure 11. TRPS-assisted CO_2 parallel flooding characteristic curves, (a) Injection pressure curves, (b) fractional flow rate curve.

3.5. EOR Effects of TRPS

Figure 12 shows the produced fluids and oil recovery curves of each stage for water-assisted CO_2 flooding and TRPS-assisted CO_2 flooding. CO_2 can dissolve in crude oil and cause it to expand, reducing its viscosity [40,41], and has a higher recovery rate than air and N_2. The oil recovery after CO_2 flooding to gas channeling is about 46%, and liquid-assisted CO_2 flooding can effectively control the gas channeling and further improve the oil recovery. Whether it is water or TRPS solution, the enhanced oil recovery of each round of the liquid injection stage is higher than that of the CO_2 injection stage of the same round. TRPS-assisted CO_2 flooding enhances oil recovery mainly in rounds 1 and 2, and the EOR effects are 20.71% and 10.00%, respectively. This also led to poorer EOR effects in subsequent rounds, in which the EOR in the third round was lower than that of water-assisted CO_2 flooding. The main effect period of water-assisted CO_2 flooding is in the first three rounds, but the overall EOR effect is 8.21% lower than that of TRPS-assisted

CO_2 flooding. The ultimate recovery of TRPS-assisted CO_2 flooding and water-assisted CO_2 flooding can reach 78.93% and 71.07%, respectively.

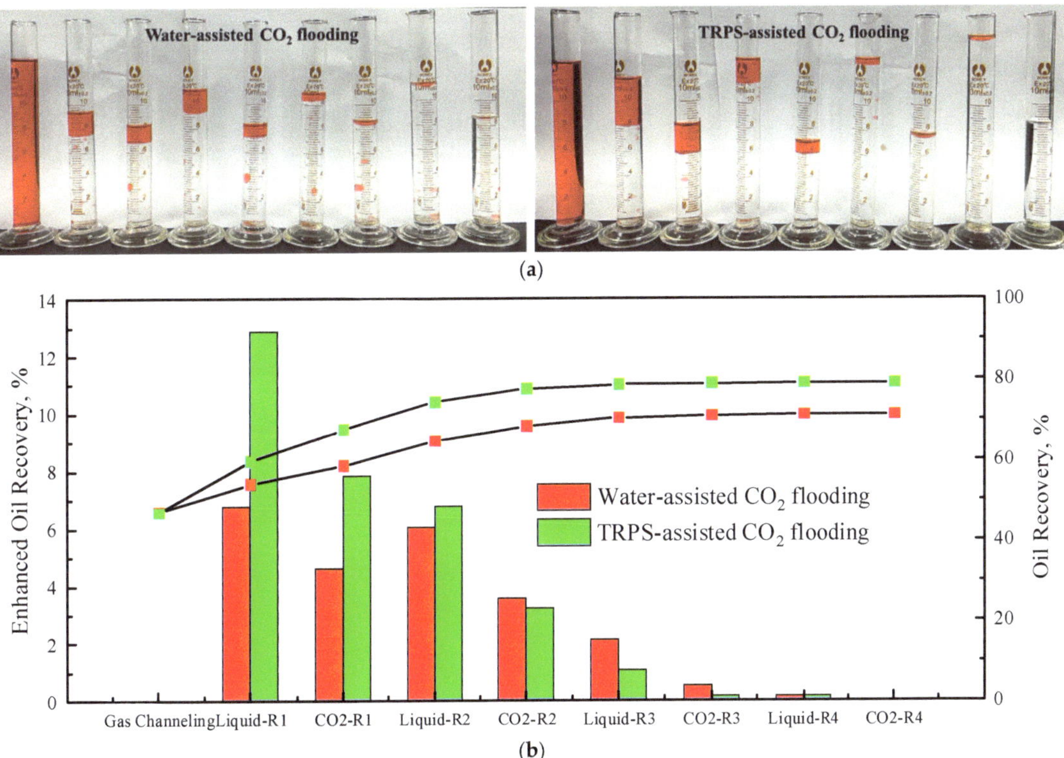

Figure 12. Experimental photos and data analysis of water-assisted CO_2 flooding and TRPS-assisted CO_2 flooding. (**a**) Liquid production photos, (**b**) EOR curves of each round.

Appendix A Table A1 compares the EOR effects of different CO_2 enhancement injections. It can be found that compared with the ultrasonic physical method and another polymer-assisted CO_2 displacement, the TRPS used in this paper has a significantly better EOR effect. In the literature review, only two scenarios obtained higher EOR than this paper (the bolded items in Appendix A Table A1). The first scenario is for heavy oil reservoirs, where CO_2 or WAG recovery is very low due to the extremely unfavorable mobility ratio, so the effect of polymer-assisted CO_2 displacement is remarkable. The second scenario is gel particle-assisted CO_2 flooding, and the research target is parallel cores. Because the profile improvement effect of gel particles is stronger than that of polymers, its EOR effect in heterogeneous reservoirs is remarkable. Therefore, the TRPS-assisted CO_2 flooding in this paper has a considerable EOR effect of 32.93% compared with CO_2 flooding and 8.21% compared with water-assisted CO_2 flooding.

Figure 13 is the pressure difference comparison curve of water-assisted CO_2 flooding and TRPS-assisted CO_2 flooding. The increase in injection pressure during the liquid injection is more significant than that of the gas injection, which can also explain the better EOR effect of each round of liquid injection in Figure 12. For water-assisted CO_2 flooding, the gas-injection pressure in each round is close to the initial gas channeling pressure, indicating that gas–water alternation can increase the flow resistance of two-phase flow to a certain extent, but the gas channeling will quickly occur and form a dominant channel again. However, the CO_2 injection pressure of each round of TRPS-assisted CO_2 flooding is significantly higher than that of water-assisted CO_2 flooding because the molecular aggregates produced by TRPS can effectively increase the flow resistance and adsorption time in the channeling channel. Figure 13 shows that, whether water or TRPS solution, the injection pressure of each round of the water injection stage and the end of the CO_2 injection stage is significantly higher than that of the previous round, which has a significant cumulative effect. The higher injection pressure is the main reason for the quicker effect of TRPS-assisted CO_2 flooding and better EOR effect. The injection pressure of liquid-assisted CO_2 flooding will continue to increase with the injection round, but the EOR will drop significantly after the third round, so the optimal development round should be three. In addition, the increased injection pressure difference of liquid-assisted CO_2 displacement can not only expand the swept volume [42], but also effectively promote the dissolution of CO_2 into crude oil and reduce its viscosity. The high-pressure CO_2 environment will reduce the viscosity of the TRPS solution, but the viscosity of crude oil will be reduced to a greater extent, so TRPS-assisted CO_2 flooding has good EOR potential.

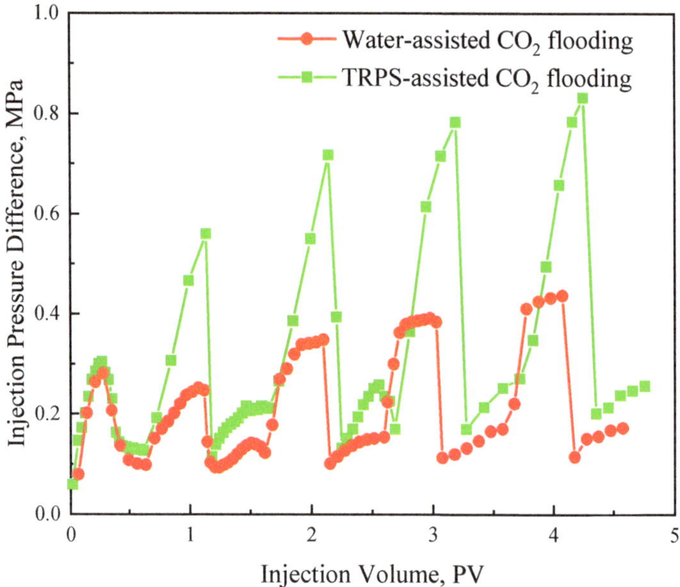

Figure 13. The injection pressure difference curves of water-assisted CO_2 flooding and TRPS-assisted CO_2 flooding.

4. Conclusions

This paper investigated the improvement of CO_2 channeling in low-permeability reservoirs by polymer solutions. Through the experimental evaluation of the static properties, flow properties, resistance-increasing effect, and EOR effect of the polymer solution, the polymer surfactant (TRPS) with CO_2 response was selected as the assisted agent, which has a good application effect. The specific conclusions are as follows:

(1) TRPS has a certain viscosity-increasing property, and the viscosity of the solution is 4.57 cP when the concentration is 1000 mg/L. It has good temperature resistance, and its viscosity retention rate is above 80% after aging at 80 °C for 30 days. In addition, TRPS can react with CO_2 to increase the size of molecular aggregates significantly, and the viscosity retention rate is about 120% after aging in a CO_2 environment for 10 days.

(2)   The R_F and R_{FF} of the TRPS solution with a concentration of 1000 mg/L in 5 mD cores are 1.36 and 1.28, respectively, showing good injectivity performance. Increasing the concentration of TRPS to 1500 mg/L had little effect on its injectivity performance.

(3)   The injection pressure of TRPS and CO_2 co-injection is between the injection of HPAM with the same viscosity and the injection of TRPS solution alone, which has good flow performance and resistance-increasing effect. The η and η' of TRPS-assisted CO_2 flooding increase with increased permeability, concentration of TRPS solution, and injection rounds. When the permeability is 5 mD, the base pressure of gas channeling is high, which will reduce the matching system between TRPS and the reservoir, thus affecting the change law of η'.

(4)   The effect of TRPS solution on profile improvement: 1000 mg/L TPRS + CO_2 > 1000 mg/L TRPS > 500 mg/L TRPS + CO_2, considering the η' and profile improvement effect, the application concentration of TRPS should be 1000 mg/L.

(5)   The EOR effect of TPRS-assisted CO_2 flooding is 8.21% higher than that of water-assisted CO_2 flooding. The EOR effect of TRPS-assisted CO_2 displacement is mainly reflected in the first to second rounds, while the EOR effect of water-assisted CO_2 displacement is primarily reflected in the first to third rounds. The injection pressure of liquid-assisted CO_2 flooding has a cumulative impact of multiple rounds, so the optimal injection round is 3.

Author Contributions: Methodology, X.C.; Investigation, X.C.; Resources, X.S.; Data curation, Z.L.; Writing—original draft, X.C.; Writing—review & editing, Z.L. and J.L.; Supervision, Y.L. and S.L.; Project administration, Y.L. and S.L. All authors have read and agreed to the published version of the manuscript.

Funding: The work was supported by the National Natural Science Foundation of China (Grant 52074318, 52174032, 52304035), the China Scholarship Council (No. 202106440061).

Institutional Review Board Statement: Not applicable.

Data Availability Statement: Not applicable.

Acknowledgments: This work was supported by the National Natural Science Foundation of China (Grant 52074318, 52174032, and 52304035) and the China Scholarship Council (No. 202106440061). The authors express their appreciation to technical reviewers for their constructive comments.

Conflicts of Interest: Author Xin Chen was employed by the company Xi'an Shiyou University. The remaining authors declare that the research was conducted in the absence of any commercial or financial relationships that could be construed as a potential conflict of interest.

Appendix A

Table A1. EOR comparison of CO_2 flooding with different strengthening methods.

Number	Literature	Contents	Agent	Models	EOR
0	This paper	Polymer-assisted CO_2 flooding	CO_2-responsive polymer, TRPS	Cores, 5–20 mD	Based on CO_2 flooding is 32.93%; based on water-assisted CO_2 flooding is 8.21%
1	Chaturvedi [43]	Polymer-enhanced carbonated water injection	PAM	Sandpack, ~780 mD	Based on water flooding is 16.00%
2	Zhao [44]	CO_2 foam	AOS	Microfludic	Based on water flooding is 13.8–22.4%
3	Hossein [45]	Ultrasound-assisted CO_2 flooding	/	Sandpack	Based on CO_2 flooding is less than 15%

Table A1. *Cont.*

Number	Literature	Contents	Agent	Models	EOR
4	Li [42]	Polymer-assisted CO_2 flooding	Polymer	Simulation, 1–2000 mD	Based on water flooding ranges from 20–30%
5	Luo [33]	Polymer-assisted CO_2 flooding	Thermo- and CO_2-triggered copolymer	Core	Based on water flooding is 21–23%
6	Yang [46]	Polymer-assisted CO_2 flooding in heavy oil reservoir	Polymer	Simulation, 500 mD	**Based on WAG is 57%**
7	Gandomkar [47]	Polymer thickening CO_2 flooding	Polydimethylsiloxane (PDMS)	Core, 6–8 mD	Based on CO_2 flooding is 6–15%
8	Manan [48]	Polymer/surfactant/ nano-particles assisted CO_2 flooding	AOS, NPs (TiO_2, CuO, SiO_2, and Al_2O_3)	Sandpack	Based on water flooding is about 5.1–15.6%
9	Zaberi [49]	Polymer-assisted CO_2 flooding	Polyfluoroacrylate (PFA)	Berea sandstone, ~31 mD	Based on CO_2 flooding is 16%
10	Liu [5]	Microgel alternate CO_2	Microgel	Core, 2800 mD/ 780 mD/360 mD	**Based on WAG is 12.4%**

References

1. Kang, W.-L.; Zhou, B.-B.; Issakhov, M.; Gabdullin, M. Advances in enhanced oil recovery technologies for low permeability reservoirs. *Pet. Sci.* **2022**, *19*, 1622–1640. [CrossRef]
2. Chen, Z.; Su, Y.-L.; Li, L.; Meng, F.-K.; Zhou, X.-M. Characteristics and mechanisms of supercritical CO_2 flooding under different factors in low-permeability reservoirs. *Pet. Sci.* **2022**, *19*, 1174–1184. [CrossRef]
3. Zhao, F.; Wang, P.; Huang, S.; Hao, H.; Zhang, M.; Lu, G. Performance and applicable limits of multi-stage gas channeling control system for CO_2 flooding in ultra-low permeability reservoirs. *J. Pet. Sci. Eng.* **2020**, *192*, 107336. [CrossRef]
4. Liu, F.; Yue, P.; Wang, Q.; Yu, G.; Zhou, J.; Wang, X.; Fang, Q.; Li, X. Experimental Study of Oil Displacement and Gas Channeling during CO_2 Flooding in Ultra—Low Permeability Oil Reservoir. *Energies* **2022**, *15*, 5119. [CrossRef]
5. Liu, Z.; Zhang, J.; Li, X.; Xu, C.; Chen, X.; Zhang, B.; Zhao, G.; Zhang, H.; Li, Y. Conformance control by a microgel in a multi-layered heterogeneous reservoir during CO_2 enhanced oil recovery process. *Chin. J. Chem. Eng.* **2022**, *43*, 324–334. [CrossRef]
6. Youyi, Z.; Qingfeng, H.; Guoqing, J.; Desheng, M.A.; Zhe, W.A.N.G. Current development and application of chemical combination flooding technique. *Petroleum Explor. Develop.* **2013**, *40*, 96–103.
7. Wang, D.; Zhang, Z.; Cheng, J.; Yang, J.; Gao, S.; Li, L. Pilot Tests of Alkaline/Surfactant/Polymer Flooding in Daqing Oil Field. *SPE Reserv. Eng.* **1997**, *12*, 229–233. [CrossRef]
8. Yupu, W.; He, L. Commercial Success of Polymer Flooding in Daqing Oilfield—Lessons Learned. In Proceedings of the SPE Asia Pacific Oil and Gas Conference and Exhibition, Adelaide, Australia, 11 September 2006. [CrossRef]
9. Mishra, S.; Bera, A.; Mandal, A. Effect of Polymer Adsorption on Permeability Reduction in Enhanced Oil Recovery. *J. Pet. Eng.* **2014**, *2014*, 1–9. [CrossRef]
10. Kumar, S.; Mandal, A. Rheological properties and performance evaluation of synthesized anionic polymeric surfactant for its application in enhanced oil recovery. *Polymer* **2017**, *120*, 30–42. [CrossRef]
11. Verma, J.; Mandal, A. Potential effective criteria for selection of polymer in enhanced oil recovery. *Pet. Sci. Technol.* **2021**, *40*, 879–892. [CrossRef]
12. Guangzhi, L.I.A.O.; Qiang, W.A.N.G.; Hongzhuang, W.A.N.G.; Weidong, L.; Zhengmao, W. Chemical flooding development status and prospect. *Acta Petrolei Sin.* **2017**, *38*, 196.
13. Li, X.; Zhang, F.; Liu, G. Review on polymer flooding technology. In *IOP Conference Series: Earth and Environmental Science*; IOP Publishing: Xiamen, China, 2021; p. 012199.
14. Liu, J.; Zhong, L.; Wang, C.; Li, S.; Yuan, X.; Liu, Y.; Meng, X.; Zou, J.; Wang, Q. Investigation of a high temperature gel system for application in saline oil and gas reservoirs for profile modification. *J. Pet. Sci. Eng.* **2020**, *195*, 107852. [CrossRef]
15. Chen, X.; Li, Y.; Liu, Z.; Zhang, J.; Trivedi, J.; Li, X. Experimental and theoretical investigation of the migration and plugging of the particle in porous media based on elastic properties. *Fuel* **2023**, *332*, 126224. [CrossRef]
16. Chen, X.; Li, Y.; Liu, Z.; Zhang, J.; Chen, C.; Ma, M. Investigation on matching relationship and plugging mechanism of self-adaptive micro-gel (SMG) as a profile control and oil displacement agent. *Powder Technol.* **2020**, *364*, 774–784. [CrossRef]

17. Chen, X.; Li, Y.; Liu, Z.; Li, X.; Zhang, J.; Zhang, H. Core- and pore-scale investigation on the migration and plugging of polymer microspheres in a heterogeneous porous media. *J. Pet. Sci. Eng.* **2020**, *195*, 107636. [CrossRef]

18. Chen, X.; Li, Y.-Q.; Liu, Z.-Y.; Trivedi, J.; Gao, W.-B.; Sui, M.-Y. Experimental investigation on the enhanced oil recovery efficiency of polymeric surfactant: Matching relationship with core and emulsification ability. *Pet. Sci.* **2023**, *20*, 619–635. [CrossRef]

19. Zhong, L.; Liu, J.; Yuan, X.; Wang, C.; Teng, L.; Zhang, S.; Wu, F.; Shen, W.; Jiang, C. Subsurface Sludge Sequestration in Cyclic Steam Stimulated Heavy-Oil Reservoir in Liaohe Oil Field. *SPE J.* **2020**, *25*, 1113–1127. [CrossRef]

20. Hu, J.; Li, A.; Memon, A. Experimental Investigation of Polymer Enhanced Oil Recovery under Different Injection Modes. *ACS Omega* **2020**, *5*, 31069–31075. [CrossRef]

21. Liang, X.; Shi, L.; Cheng, L.; Wang, X.; Ye, Z. Optimization of polymer mobility control for enhanced heavy oil recovery: Based on response surface method. *J. Pet. Sci. Eng.* **2021**, *206*, 109065. [CrossRef]

22. Dann, M.W.; Burnett, D.B.; Hall, L.M. Polymer performance in low permeability reservoirs. In Proceedings of the SPE International Conference on Oilfield Chemistry, Dallas, TX, USA, 25 January 1982.

23. Ghosh, P.; Sharma, H.; Mohanty, K.K. ASP flooding in tight carbonate rocks. *Fuel* **2018**, *241*, 653–668. [CrossRef]

24. Marliere, C.; Wartenberg, N.; Fleury, M.; Tabary, R.; Dalmazzone, C.; Delamaide, E. Oil Recovery in Low Permeability Sandstone Reservoirs Using Surfactant-Polymer Flooding. In Proceedings of the SPE Latin America and Caribbean Petroleum Engineering Conference, Quito, Ecuador, 19 November 2015; p. D031S030R004. [CrossRef]

25. Bennetzen, M.V.; Gilani, S.F.; Mogensen, K.; Ghozali, M.; Bounoua, N. Successful polymer flooding of low-permeability, oil-wet, carbonate reservoir cores. In Proceedings of the Abu Dhabi International Petroleum Exhibition and Conference, Abu Dhabi, United Arab Emirates, 10–13 November 2014; p. D021S019R004.

26. Al-Murayri, M.T.; Kamal, D.S.; Al-Sabah, H.M.; AbdulSalam, T.; Al-Shamali, A.; Quttainah, R.; Glushko, D.; Britton, C.; Delshad, M.; Liyanage, J.; et al. Low-Salinity Polymer Flooding in a High-Temperature Low-Permeability Carbonate Reservoir in West Kuwait. In Proceedings of the SPE Kuwait Oil and Gas Show and Conference, Mishref, Kuwait, 13–16 October 2019. [CrossRef]

27. Leon, J.M.; Castillo, A.F.; Perez, R.; Jimenez, J.A.; Izadi, M.; Mendez, A.; Castillo, O.P.; Londoño, F.W.; Zapata, J.F.; Chaparro, C.H. A Successful Polymer Flood Pilot at Palogrande-Cebu, A Low Permeability Reservoir in the Upper Magdalena Valley, Colombia. In Proceedings of the SPE Improved Oil Recovery Conference, Oklahoma City, OK, USA, 14–18 April 2018. [CrossRef]

28. Musevic, I.; Skarabot, M.; Tkalec, U.; Ravnik, M.; Zumer, S. Two-Dimensional Nematic Colloidal Crystals Self-Assembled by Topological Defects. *Science* **2006**, *313*, 954–958. [CrossRef]

29. Su, X.; Cunningham, M.F.; Jessop, P.G. Switchable viscosity triggered by CO_2 using smart worm-like micelles. *Chem. Commun.* **2013**, *49*, 2655–2657. [CrossRef] [PubMed]

30. Lin, S.; Theato, P. CO_2-responsive polymers. *Macromol. Rapid Commun.* **2013**, *34*, 1118–1133. [CrossRef] [PubMed]

31. Zhang, Y.; Feng, Y.; Wang, J.; He, S.; Guo, Z.; Chu, Z.; Dreiss, C.A. CO_2-switchable wormlike micelles. *Chem. Commun.* **2013**, *49*, 4902–4904. [CrossRef] [PubMed]

32. Zhang, Y.; Feng, Y.; Wang, Y.; Li, X. CO_2-Switchable Viscoelastic Fluids Based on a Pseudogemini Surfactant. *Langmuir* **2013**, *29*, 4187–4192. [CrossRef] [PubMed]

33. Luo, X.; Zheng, P.; Gao, K.; Wei, B.; Feng, Y. Thermo- and CO_2-triggered viscosifying of aqueous copolymer solutions for gas channeling control during water-alternating-CO_2 flooding. *Fuel* **2021**, *291*, 120171. [CrossRef]

34. Shen, H.; Yang, Z.; Li, X.; Peng, Y.; Lin, M.; Zhang, J.; Dong, Z. CO_2-responsive agent for restraining gas channeling during CO_2 flooding in low permeability reservoirs. *Fuel* **2021**, *292*, 120306. [CrossRef]

35. Wu, Y.; Liu, Q.; Liu, D.; Cao, X.P.; Yuan, B.; Zhao, M. CO_2 responsive expansion hydrogels with programmable swelling for in-depth CO_2 conformance control in porous media. *Fuel* **2023**, *332*, 126047. [CrossRef]

36. Zhang, D.; Li, Y.; Bao, Z. A laboratory experimental study of feasibility of polymer flood for middle-low permeability reservoirs in Daqing. *Oilfield Chem.* **2005**, *22*, 78–80.

37. Pal, N.; Zhang, X.; Ali, M.; Mandal, A.; Hoteit, H. Carbon dioxide thickening: A review of technological aspects, advances and challenges for oilfield application. *Fuel* **2022**, *315*, 122947. [CrossRef]

38. Yang, J.; Dong, H. CO_2-responsive aliphatic tertiary amine-modified alginate and its application as a switchable surfactant. *Carbohydr. Polym.* **2016**, *153*, 1–6. [CrossRef] [PubMed]

39. Fan, W.; Tong, X.; Farnia, F.; Yu, B.; Zhao, Y. CO_2-Responsive Polymer Single-Chain Nanoparticles and Self-Assembly for Gas-Tunable Nanoreactors. *Chem. Mater.* **2017**, *29*, 5693–5701. [CrossRef]

40. Kumar, S.; Mandal, A. A comprehensive review on chemically enhanced water alternating gas/CO_2 (CEWAG) injection for enhanced oil recovery. *J. Pet. Sci. Eng.* **2017**, *157*, 696–715. [CrossRef]

41. Kumar, N.; Sampaio, M.A.; Ojha, K.; Hoteit, H.; Mandal, A. Fundamental aspects, mechanisms and emerging possibilities of CO_2 miscible flooding in enhanced oil recovery: A review. *Fuel* **2022**, *330*, 125633. [CrossRef]

42. Li, H.; Zheng, S.; Yang, D. Enhanced swelling effect and viscosity reduction of solvent (s)/CO_2/heavy-oil systems. *Spe J.* **2013**, *18*, 695–707. [CrossRef]

43. Chaturvedi, K.R.; Trivedi, J.; Sharma, T. Evaluation of Polymer-Assisted Carbonated Water Injection in Sandstone Reservoir: Absorption Kinetics, Rheology, and Oil Recovery Results. *Energy Fuels* **2019**, *33*, 5438–5451. [CrossRef]

44. Zhao, J.; Torabi, F.; Yang, J. The synergistic role of silica nanoparticle and anionic surfactant on the static and dynamic CO_2 foam stability for enhanced heavy oil recovery: An experimental study. *Fuel* **2020**, *287*, 119443. [CrossRef]

45. Hamidi, H.; Haddad, A.S.; Mohammadian, E.; Rafati, R.; Azdarpour, A.; Ghahri, P.; Ombewa, P.; Neuert, T.; Zink, A. Ultrasound-assisted CO_2 flooding to improve oil recovery. *Ultrason. Sonochem.* **2017**, *35*, 243–250. [CrossRef]
46. Yang, Y.; Li, W.; Zhou, T.; Dong, Z. Using Polymer Alternating Gas to Enhance Oil Recovery in Heavy Oil. *IOP Conf. Ser. Earth Environ. Sci.* **2018**, *113*, 012182. [CrossRef]
47. Gandomkar, A.; Torabi, F.; Riazi, M. CO_2 mobility control by small molecule thickeners during secondary and tertiary enhanced oil recovery. *Can. J. Chem. Eng.* **2020**, *99*, 1352–1362. [CrossRef]
48. Manan, M.A.; Farad, S.; Piroozian, A.; Esmail, M.J.A. Effects of Nanoparticle Types on Carbon Dioxide Foam Flooding in Enhanced Oil Recovery. *Pet. Sci. Technol.* **2015**, *33*, 1286–1294. [CrossRef]
49. Zaberi, H.A.; Lee, J.J.; Enick, R.M.; Beckman, E.J.; Cummings, S.D.; Dailey, C.; Vasilache, M. An experimental feasibility study on the use of CO_2-soluble polyfluoroacrylates for CO_2 mobility and conformance control applications. *J. Pet. Sci. Eng.* **2019**, *184*, 106556. [CrossRef]

Article

Application of CO$_2$-Soluble Polymer-Based Blowing Agent to Improve Supercritical CO$_2$ Replacement in Low-Permeability Fractured Reservoirs

Mingxi Liu [1], Kaoping Song [1,*], Longxin Wang [1], Hong Fu [1] and Jiayi Zhu [2]

1. Unconventional Petroleum Research Institute, China University of Petroleum, Beijing 102249, China; 1995liumingxi@sina.com (M.L.); 15717673116@163.com (L.W.); catherinefu90@sina.com (H.F.)
2. College of Petroleum Engineering, Xi'an Shiyou University, Xi'an 710065, China; 18813109909@163.com
* Correspondence: skpskp01@sina.com; Tel.: +86-1380465350

Abstract: Since reservoirs with permeability less than 10 mD are characterized by high injection difficulty, high-pressure drop loss, and low pore throat mobilization during the water drive process, CO$_2$ is often used for development in actual production to reduce the injection difficulty and carbon emission simultaneously. However, microfractures are usually developed in low-permeability reservoirs, which further reduces the injection difficulty of the driving medium. At the same time, this makes the injected gas flow very fast, while the gas utilization rate is low, resulting in a low degree of recovery. This paper conducted a series of studies on the displacement effect of CO$_2$-soluble foaming systems in low-permeability fractured reservoirs (the permeability of the core matrix is about 0.25 mD). For the two CO$_2$-soluble blowing agents CG-1 and CG-2, the effects of the CO$_2$ phase state, water content, and oil content on static foaming performance were first investigated; then, a more effective blowing agent was preferred for the replacement experiments according to the foaming results; and finally, the effects of the blowing agents on sealing and improving the recovery degree of a fully open fractured core were investigated at different injection rates and concentrations, and the injection parameters were optimized. The results show that CG-1 still has good foaming performance under low water volume and various oil contents and can be used in subsequent fractured core replacement experiments. After selecting the injection rate and concentration, the blowing agent can be used in subsequent fractured cores under injection conditions of 0.6 mL/min and 2.80%. In injection conditions, the foaming agent can achieve an 83.7% blocking rate and improve the extraction degree by 12.02%. The research content of this paper can provide data support for the application effect of a CO$_2$-soluble blowing agent in a fractured core.

Keywords: fractured core; supercritical CO$_2$; CO$_2$-soluble blowing agent; static foaming performance evaluation; blocking rate; enhanced recovery; injection parameter optimization

Citation: Liu, M.; Song, K.; Wang, L.; Fu, H.; Zhu, J. Application of CO$_2$-Soluble Polymer-Based Blowing Agent to Improve Supercritical CO$_2$ Replacement in Low-Permeability Fractured Reservoirs. *Polymers* **2024**, *16*, 2191. https://doi.org/10.3390/polym16152191

Academic Editor: Marcelo Antunes

Received: 5 July 2024
Revised: 26 July 2024
Accepted: 27 July 2024
Published: 1 August 2024

1. Introduction

With the exploitation of conventional oil reservoirs gradually entering the bottleneck period, many old oil fields have progressively entered the stage of high water content and low extraction degree. Unconventional reservoirs have gradually become the focus of petroleum exploration and development, in which the efficient exploitation of low-permeability reservoirs with permeability less than 50 mD is the current key research direction. In this type of reservoir, because of its small reservoir permeability, the water injection process pressure loss is large, and an imbalance of injection and production often occurs, so the conventional water drive development effect is poor. This type of reservoir is usually extracted by fracturing to produce artificial fractures. Although the existence of cracks exacerbates the flow of oil repellents, it also effectively reduces the differential pressure of the drive, dramatically reduces the energy loss, and helps transport the oil repellents to the deep part of the reservoir [1]. For reservoirs with a 10–50 mD permeability, combined with fracturing and reasonable well network deployment, conventional water drive can effectively utilize the crude oil in the reservoir [2,3]. However, for reservoirs with permeability less than 10 mD, the pore size of the core is too small. Under the combined influence of capillary force and oil–water interfacial tension, the water needs higher pressure to enter the interior of the pore throat, and the seepage resistance is higher. However, in a fracture–pore seepage system, water flows along the fracture surface where the seepage resistance is extremely low, and the proportion of water that can enter the interior of the pore throat for displacement is extremely low, making water drive utilization of fracture surface crude oil challenging. Compared with water, supercritical CO_2 has extremely low viscosity and strong diffusivity, which can diffuse into core matrix pores and extract light components, so CO_2 drive can effectively develop such reservoirs.

For reservoirs with less fracture development, fractures can be created by fracturing. The existence of artificial fractures dramatically improves the contact area between CO_2 and rock. This facilitates the transport of CO_2 to the deeper part of the reservoir, achieving better development results. The presence of fractures significantly increases the CO_2 wave volume. CO_2 expands the pore-throat utilization interval while significantly reducing the lower limit of pore-throat utilization. The presence of both simultaneously improves the degree of recovery in low-permeability reservoirs [4–8]. Currently, scholars believe that CO_2 in fractures mainly diffuses in two modes of movement as follows: CO_2 displaces crude oil in fractures while moving crude oil in the matrix through the swelling and extraction of crude oil [9]. As the permeability of the matrix increases, the diffusion phenomenon becomes more evident. Previous experiments found that CO_2 can move part of the 30–100 nm tiny pore throats. If the proportion of <10 nm pore throats in some cores is higher, the diffusion effect of CO_2 is seriously weakened, and it is more difficult to move [10]. Although CO_2 has better dissolution and extraction properties, the fracture morphology directly determines the CO_2 wave area, which directly affects the final extraction degree of the CO_2 drive. In the actual development process, it is common to use the fracturing of the seam network to expand the contact area between CO_2 and the fracture surface [11,12]. However, the existence of cracks will produce severe gas flushing, which will not only make the gas–oil ratio at the extraction end unpredictable but also make the reservoir pressure drop rapidly [13]; therefore, the gas flushing control of fractured reservoirs is also significant in actual production.

In order to solve the problem of gas flushing, scholars first proposed the alternating water–gas CO_2 injection process. This method effectively maintains the reservoir pressure, facilitates the diffusion of CO_2 into the matrix, and has a specific effect on controlling gas flushing in medium-high permeability reservoirs above 50 mD [14,15]. In order to control gas flushing in fractured reservoirs more effectively, scholars have successively developed blocking agents such as gels and microspheres [16,17]. For example, Yakai Li et al. developed a new type of high-temperature slow-release gel, which improved the recovery degree by 20% in a core with matrix permeability of 20 mD and fracture opening of 150 μm [18]. Haizhuang Jiang et al. used a new type of functional polymer to block a core with a matrix permeability of 20 mD and fracture permeability of 10,000 mD, which improved the recovery degree by 23.63% in the late stage of the CO_2 drive [19]. Chengli Zhang et al. developed a salt-resistant polymer–surfactant synergistic composite low interfacial tension foam system, which increased the development degree by 13.17% after blocking the low-permeability reservoir of 10 mD [20]. Weiyao Zhu et al. developed a novel nano-microsphere system, which temporarily sealed the fractured core through the bridge and retention of the microsphere particles [21].

Moreover, retention temporarily seals the cracks and controls the oil repellent's whole flow field, which regulates the gas flow in the subsequent gas drive process. With the advancement of technology, scholars have developed new blocking agents such as CO_2 corresponding gel and CO_2 corrosion-resistant microsphere particles, which are more targeted and have achieved good effects in blocking and improving the degree of recovery in CO_2 drive experiments in fractured cores with permeability more significant than 1 mD [21,22]. For reservoirs with widely developed fractures at various scales, scholars injected gels, polymers, microsphere particles, and other agents sequentially to achieve the sequential blocking of fractures from large to small, and with the implementation of alternating water and gas injection and other implementation processes, they also achieved better blocking and improved recovery [8,23,24].

However, later studies found that in low-permeability fractured reservoirs with rock matrix permeability less than 1 mD, the presence of water in the fracture is detrimental to the degree of CO_2 recovery, and the higher water saturation in the fracture reduces the ability of CO_2 diffusion into the matrix; therefore, water acts as an inhibitor to CO_2 drive [25,26]. Xu Li et al. developed a new water-soluble thixotropic gel system, which had good stability and good sealing performance in the matrix permeability of 0.96 mD and fracture openings of 15 mm fractured rock cores, but the gas drive did not improve the degree of extraction after sealing [27]. Qipeng Ma et al. measured the relative permeability curve of the nano-microsphere system in a low-permeability homogeneous core and experimentally found that in a core with a permeability greater than 10 mD, the dispersed system could effectively expand the two-phase oil–water flow. However, in cores with permeability less than 10 mD, the system reduced the water-phase permeability but did not improve the degree of recovery [28]. Daijun Du et al. used microsphere particles to seal fractured cores with a matrix permeability of 0.1–1 mD and a fracture permeability of 5000 mD; they also found that microspheres had an excellent blocking performance for a fracture after the experiments. However, the subsequent replacement was ineffective in increasing the degree of extraction [29]. This indicates that a fractured core with matrix permeability of less than 1 mD is not suitable for water drive development, and it is also not suitable for use in water-mediated agents for plugging.

In order to increase the degree of matrix development while fracture sealing in extra-low permeability fractured cores (matrix permeability less than 1 mD), some scholars have used a direct injection of CO_2 foam for the replacement. From the macroscopic point of view, the presence of foam increases the injection pressure and promotes the diffusion ratio of the replacement medium from the fracture matrix while making the gas flushing rate decrease [30]; from the microscopic point of view, the injected foam forms resistance through the Jamin effect, which allows the replacement medium to drive off the crude oil within the pores of the fracture face [31]. In practice, factors such as the roughness of the fracture surface, injection flow rate, agent concentration, and total injected volume affect the sealing performance of the foam and must be evaluated comprehensively [32]. The stability of surfactant foaming alone is sometimes not very good, which reduces the blocking performance. Methods such as mixing CO_2 and N_2 foaming and adding nanoparticles to strengthen the surface properties of the foam have achieved good results in core replacement experiments [33–36].

Although the direct injection of foam can simultaneously reduce the rate of gas flushing and increase the degree of core recovery, the injectability of the system is relatively poor, resulting in higher injection pressure, higher energy loss, a smaller range of foam recovery, and other unfavorable phenomena [37]. In order to solve this problem, scholars developed a foaming agent that can be dissolved in both water and CO_2, compared with the traditional water-soluble surfactant, gas–water amphiphilic surfactant dissolved in CO_2 and injected into the core, and the foam formed with the water inside the fracture surface, significantly accelerating the expansion rate and area of the agent. The injection capacity is greatly improved [38] through continuous CO_2 injection in the leading edge of the foam, which can be generated at the leading edge of the exfoliation by successive CO_2 injections. Even if defoamed, foam can be generated again with water under subsequent CO_2 injections [39]. The solubility of the agent in water and CO_2 is different, and high solubility in water enhances adsorption during transport [40], so alcohols are often used as co-solvents to improve solubility in CO_2 in experiments. Alcohols significantly increase the interaction between CO_2 and polymer molecular chains, with ethanol having the most significant effect [41]. Indoor experiments showed that adding ethanol improved the foaming ability and stability of the agent. This helped foam regeneration [42], but it should be noted that the high solubility of the agent in CO_2 can lead to poor foam stability, which is not conducive to blocking and mining. The amount of ethanol needs to be controlled [43]. Poor foam stability occurs when some blowing agents come into contact with crude oil [44]. The actual blocking performance and enhanced extraction are controlled by temperature, pressure, oil and water content, fracture morphology, agent concentration, and other factors, so the agent's effectiveness must be evaluated through actual replacement experiments [45].

With the gradual decrease in global oil resources, extra-low permeability oil and gas reservoirs with K < 1 mD have attracted much attention in recent years. Because of the pore structure, oil–water interfacial tension, and other factors, such reservoirs cannot be developed by water drive, and supercritical CO_2 has become the focus of current research because of its superior performance. To improve the wave volume of oil repellent in low-permeability reservoirs, supercritical CO_2 fracturing is currently the mainstream reservoir modification method (hydraulic fracturing is not applicable). Therefore, controlling the gas flow in the fracture after fracturing and improving the recovery of CO_2 drive is a problem that must be faced when developing such reservoirs.

In this paper, the foaming performance of two kinds of CO_2-soluble blowing agents in different CO_2 phases, different water contents, and different oil contents is investigated for an extra-low permeability reservoir with K < 1 mD. Then, a more excellent blowing agent is screened according to the foaming situation, and a reasonable preparation method for this blowing agent is determined. Finally, the screened blowing agent is used to conduct driving experiments in a fractured core to study the effects of different injection speeds and injection concentrations on fractured core blocking and improving the degree of extraction. Good results were achieved, which have good prospects for practical application.

2. Materials and Methods

2.1. Experimental Material

Core and fracture pressing methods: Fracture morphology is one of the determining factors for the pattern of displacement, and in actual reservoirs, fracture openings, smoothness, and orientation can have a direct impact on the development effect. Since rocks with microfractures could not be drilled out of the completed core column, this paper's indoor physical simulation tests used homogeneous, non-fractured, natural, low-permeability rocks to create seams after drilling and polishing the ends. Currently, the primary method to create fractures is by cutting the core or Brazilian splitting. Cutting the core directly minimizes the differences among different fractures and provides excellent controllability, but, at the same time, the surface of the fractures is exceptionally smooth, which is not conducive to foaming the system in the core. Brazilian splitting creates fractures by applying shear stress to the cores. Although it is difficult to ensure that the fractures among different cores are precisely the same since the cores come from the same stone, the rhythms and the degree of cementation are incredibly close to each other, the similarity in the fractures pressed out is exceptionally high, and the rough and uneven surface of the fractures is more conducive to foaming the agent in the formation, which is also in line with the situation in actual reservoirs. Therefore, Brazilian splitting was used to create fractures.

Fracture morphology: Since a full open fracture is best controlled during fracture with minimal variation among cores, all cores were pressed out of a full open fracture for experimentation in the later evaluation process (as shown in Figure 1).

Foaming agent, gas, experimental water, and experimental oil: Because of the low permeability of the matrix of the target core in this paper, which is about 0.25 mD, based on the results of previous studies, this type of core is not suitable for water injection development, and, likewise, it is not suitable for injecting water-soluble agents. Therefore, two high-temperature-resistant, high-salt CO_2-soluble foaming agents were selected for this study, and the foaming method in the core was adopted for sealing. The gas was high-pressure CO_2 with a purity of 99.9%, which was injected into the core after mixing the blowing agent and pressurizing to the specified pressure. In order to simulate the real stratigraphic situation and examine the actual blocking performance of the agent, the experiment was carried out using mineralized water with mineralization of 9800 mL/L for saturated cores; the specific formulations are shown in Table 1 below. The experiment was conducted with simulated oil, and the viscosity was 2.135 mPa·s under 25 °C/0.1 MPa. (The experiment was conducted for a low-permeability fractured reservoir in the Bohai Sea, with a matrix permeability of 0.1–0.5 mD and $NaHCO_3$-type formation water, which needs to be developed twice after experiencing the pre-supercritical CO_2 drive. This resulted in a decline in reservoir temperature and pressure, severe gas flushing, and a low extraction level).

Brazilian splitting   A complete fracture (core completely cracked)

Figure 1. A Brazilian split and a fully open fracture core.

Table 1. Ionic species and concentration of experimental water.

Ionic Species	K^+ & Na^+	Mg^{2+}	Ca^{2+}	Cl^-	SO_4^{2-}	HCO_3^-	CO_3^{2-}
Ionic concentration mg/L	2704.30	11.87	68.49	880.50	20.45	4690.94	68.49
Total mineralization			8912.08 mg/L				

The inorganic salts and white oil were purchased from Shanghai Aladdin Reagent Co. (Shanghai, China). All equipment from Yongruida Technology Co. (Beijing, China). All CO_2 comes from Shunxing Co. (Beijing, China).

2.2. Experimental Methods

2.2.1. Evaluation Methods and Test Programs for the Foaming Capacity of Blowing Agents

Experimental program: The foaming performance with liquid CO_2 and supercritical CO_2 was first investigated for the two blowing agents. The liquid CO_2 foaming temperature and pressure were 22 °C and 10 MPa; the supercritical CO_2 foaming temperature and pressure were 70 °C and 20 MPa, and the water content was 10% in all cases (the volume of the foaming instrument was 250 mL, and the water volume was 25 mL). The concentration of the agent was 1.6% (CO_2 and water were solvents in the system, the volume of solvent was 250 mL, the foaming agent was 4 mL, and the concentration was 1.6%). The water content in the supercritical CO_2 foaming system was subsequently varied, with the blowing agent content remaining constant at 1.6%, and the foaming performance was investigated for water contents of 5%, 10%, and 20% (12.5, 25, and 50 mL of water, respectively). Finally, water–oil ratios of 100:3, 100:10, and 100:20 were set at 10% water content to investigate the effect of oil content on the foaming performance of the supercritical CO_2 foaming system and to conduct foaming agent screening.

A visual high-temperature and high-pressure foaming apparatus was used to evaluate the foaming performance of the agents. The experimental steps and experimental flowchart are shown in Figure 2 and described as follows:

1. First, the solution was configured in the foaming apparatus according to the concentration of the agent, water content, and oil content required by the experiment. Then, the foaming apparatus was closed and sealed, followed by vacuuming for 15 min (to avoid the loss of the foaming agent and oil, the solution was prepared directly inside the instrumentation; to avoid the influence of air, the equipment was evacuated prior to CO_2 injection);

2. Since the gas pressure will rise sharply during the heating process, to ensure the experiment's safety, do not increase the pressure before the end of the heating. If the pressure is not reached at the end of the heating process, the pressure should be increased slowly and not too quickly.

3. After the temperature and pressure in the equipment were stabilized, the stirring switch was turned on. After the foaming was stabilized, the foaming performance was examined according to the foaming height and half-life of the agent, and the agent screening was completed.

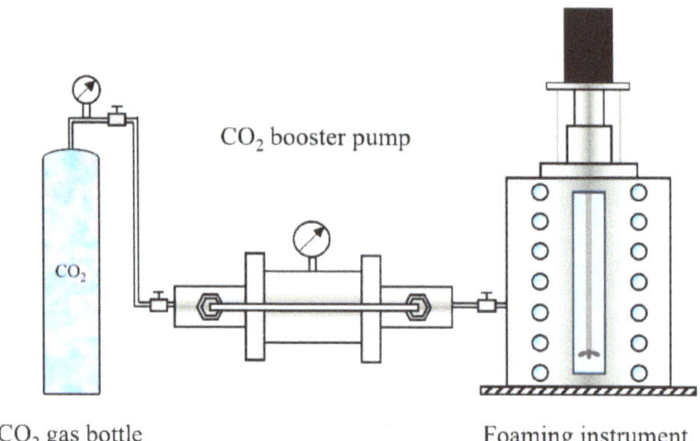

Figure 2. Flowchart of static evaluation experiment of foaming agent

2.2.2. Methods and Experimental Protocols for CO_2 + Blowing Agent Exfoliation of Fractured Cores

Experimental protocol: Prior to the sealing experiments, pure CO_2 replacement experiments (as a control test) were first carried out on fractured cores at different injection flow rates (injection rates of 0.1, 0.2, 0.4, 0.6, and 0.8 mL/min), followed by CO_2 + foaming agent replacement experiments. In order to investigate the sealing of the fractured core by the blowing agent after CO_2 replacement and the ability to increase the degree of extraction, the experiment was divided into three segment plugs with CO_2-CO_2 + blowing agent-CO_2 injection, where CO_2 was mixed with the blowing agent through a stirring piston. For the core with one fully open fracture, firstly, the driving effect of the blowing agent at different injection rates was investigated for the preferred injection rate (injection concentration of 1.11%, injection rate of 0.1, 0.2, 0.4, 0.6, and 0.8 mL/min). The sealing of the fractured core and the enhanced recovery characteristics of the different concentrations of agents were then examined at preferred injection rates (blowing agent concentrations of 0.56%, 1.11%, 1.67%, 2.28%, and 2.80%, respectively). The experimental temperature was 70 °C, controlled by a thermostat; the experimental minimum pressure was 20 MPa, controlled by a pressure-return valve; and the experimental core was 3.8 cm in diameter and 8 cm in length.

The foaming agent configuration method and experimental steps are shown in Figure 3a, and the foam formation schematic is shown in Figure 3b. The steps are described as follows:

1. The core was polished to the specified size, dried, and weighed, and then vacuumed and saturated with water. Because of the low permeability of the core, the vacuuming time needed to be more than 8 h, and the saturated formation water time needed to be more than 6 h. After the saturated water, the core was weighed, and the pore volume was calculated;

2. Because of the large cross-sectional area of the core and low permeability, saturated oil was injected with 20 MPa constant pressure, and after about 24 h, the saturated oil was finished. At this time, the oil saturation degree of the core was about 50% (it was difficult to saturate low permeability cores with oil, and the saturation was only 55% after 72 h of saturation);

3. The core is removed and immediately fractured at the end of saturated oil. A reasonable fracturing method is developed based on the fracture morphology, and the core is loaded into the gripper immediately after fracturing in preparation for deplacement (the fracturing process takes about 120 s, which is extremely short, and the pore sizes are small, the volatilisation of fluids in the core is very small, and the loss of oil and water during the fracturing process is negligible);

4. After loading the core, the inlet and outlet of the gripper were closed, and 2 MPa perimeter pressure was applied, followed by heating. To ensure that the core reached the experimental temperature, it was heated for more than 4 h. To avoid the volatilization of oil and water from the side of the core during the heating period, it was necessary to add a perimeter pressure so that the sleeve stuck close to the surface of the core. At the same time, the outlet and the inlet of the gripper were closed to prevent the oil and water vapours that evaporated from the core's end face from escaping out of the gripper;

5. After heating, a hand pump was used to pressurize the pressure return valve to 20 MPa. The gas was then pressurized using a booster pump, and when the gas in the piston was 20 MPa, the pressurization was complete. Inside the thermostat, the gas was injected into the core through a metal coil of about 6 m in length. As the gas flowed through the high-temperature coil, it was sufficient to reach the specified temperature and complete the phase transition;

6. After opening the inlet end of the gripper, the gas flow reached the core outlet immediately through the fracture, and the pressures at both ends were balanced after about 90 s. Subsequently, the injection flow rate was set, and the outlet end of the gripper was opened in preparation for the replacement (during the replacement process, the oil and gas production at any time during the stage time and the pressure data were recorded);

7. The injection was stopped after injecting about 1–4 PV. At this time, the pure CO_2 replacement experiment was finished; the thermostat was closed, and the pressure was removed. To switch to the CO_2 + blowing agent, we closed the gripper inlet and outlet after stopping the injection and recorded the pressure inside the core;

8. Current CO_2-soluble agents use water as an intermediate medium, and CO_2 can scramble the blowing agent out of the water during constant stirring. The stirred piston was therefore fitted with a propeller on one side and configured by pouring the aqueous solution directly into the piston, evacuating the air inside the piston using a vacuum pump (15 min), followed by injection and pressurization with CO_2;

9. After pressurizing to the specified pressure, the piston position was adjusted so that the propeller was in direct contact with the aqueous solution. The propeller was then opened and stirred for 20 min and left to stand for 20 min after the end of stirring. Subsequently, the core entrance and exit were opened, the flow rate was set after the pressure and gas flow rate stabilized, the CO_2 + blowing agent drive was started, and the experimental data were recorded. At this time, the piston pressure was slightly higher than the internal pressure of the core, and there was a small pressure disturbance after opening the outlet, which stabilized after about 30 s.

10. Step (9) was repeated after the end of the CO_2 + blowing agent replacement. After stabilization, the subsequent pure CO_2 replacement was turned on, and experimental data such as differential pressure, oil production, etc., were recorded. The experiment was completed after a certain amount of replacement.

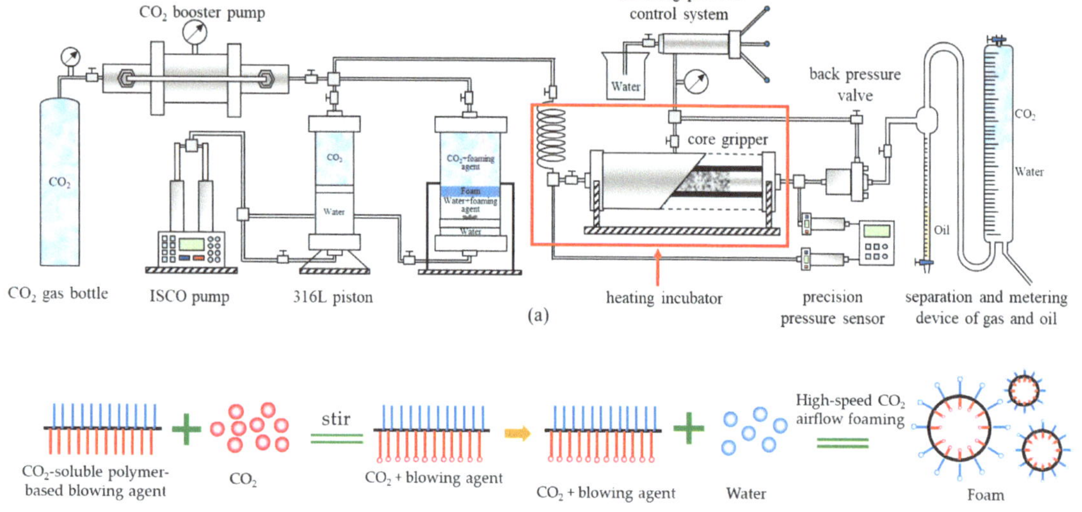

Figure 3. Flowchart of CO_2 + blowing agent drive experiments on fractured cores (**a**) and schematic diagram of foam formation (**b**).

3. Results and Discussion

3.1. Evaluation of the Foaming Capacity of CO$_2$-Blowing Agents

Using the foaming equipment, two CO_2-soluble blowing agents, CG-1 and CG-2, were selected to conduct experiments to evaluate the foaming ability under different CO_2 phases and different water and oil contents.

In the process of stirring the aqueous solution with CO_2, it does not foam immediately. In the case of the CG-1 foaming process with liquid CO_2, for example, as shown in Figure 4, at the beginning of the mixing process, foam of a large size is formed, visible to the naked eye, and spreads throughout the entire viewport during the mixing process. The foam in the middle of stirring gradually sinks to the bottom of the instrument; the upper foam is large and the lower foam is small. In late stirring, all the foam sinks to the bottom of the container, forming a stable, fine foam, and only a tiny amount of foam exists hanging on the viewport. When the foaming was over, the stirrer was closed, and the foaming height was measured. After defoaming, the half-life was measured (the stirring process of the two blowing agents was about 90–120 s).

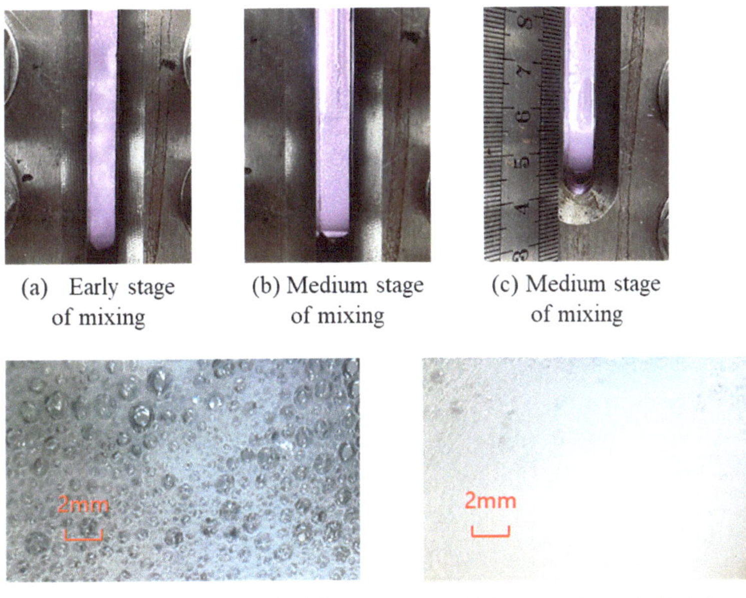

(a) Early stage of mixing   (b) Medium stage of mixing   (c) Medium stage of mixing

(d) Foam at the beginning of mixing   (e) Foam at the end of mixing

Figure 4. CG-1 and liquid CO_2 foaming process.

3.1.1. Comparison of Foaming Ability of Agents in Different Phases

Liquid CO_2 (22 °C/10 MPa): Figure 5, shows CG-1 in 10% water at a concentration of 1.6% and a foaming height of 1 cm. The foam volume is small but has excellent stability with a half-life of 450 min. For CG-2 in 10% water, a concentration of 1.6% is required to form a stable foam with a half-life of up to 1200 min, but the bubble height is still low at 2.7 cm.

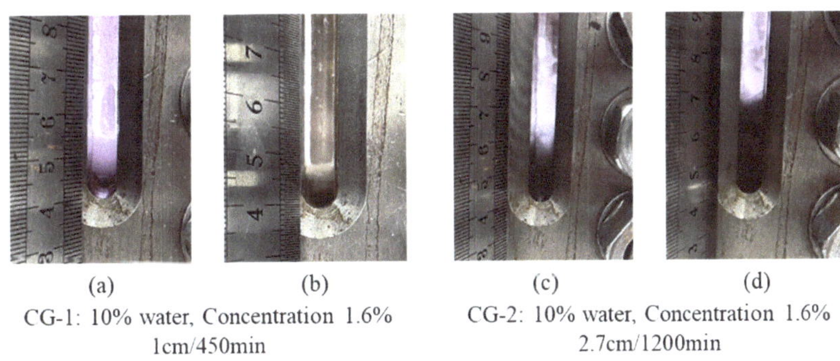

(a)   (b)   (c)   (d)

CG-1: 10% water, Concentration 1.6%   CG-2: 10% water, Concentration 1.6%
1cm/450min   2.7cm/1200min

Liquid CO_2 (22°C/10MPa)

Figure 5. Results of CG-1 and CG-2 foaming with liquid CO_2. (**a**) Photos of CG-1 at the end of foaming; (**b**) Photographs taken when CG-1 has reached its half-life; (**c**) Photos of CG-2 at the end of foaming; (**d**) Photographs of CG-2 when it has reached its half-life.

Supercritical CO_2 (70 °C, 20 MPa): Figure 6 shows CG-1 at 10% water and a concentration of 1.6% with a foaming height of 4.5 cm and a half-life of 100 min. CG-2 is shown at 10% water and a concentration of 1.6% with a foaming height of 5.3 cm and a half-life of 15 min.

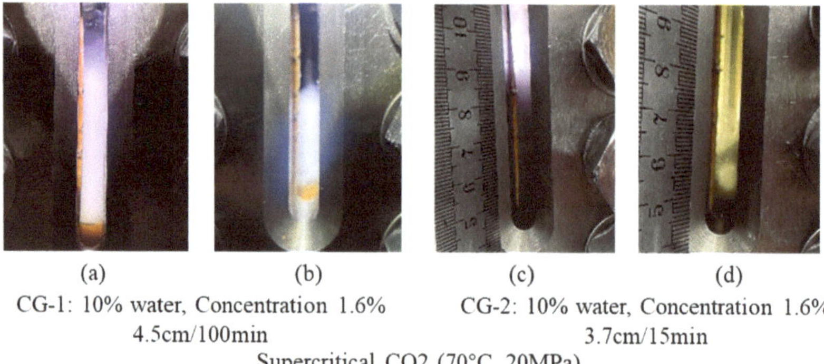

| (a) | (b) | (c) | (d) |

CG-1: 10% water, Concentration 1.6%   CG-2: 10% water, Concentration 1.6%
4.5cm/100min   3.7cm/15min

Supercritical CO2 (70°C, 20MPa)

Figure 6. Results of CG-1 and CG-2 foaming with supercritical CO_2. (**a**) Photos of CG-1 at the end of foaming; (**b**) Photographs taken when CG-1 has reached its half-life; (**c**) Photos of CG-2 at the end of foaming; (**d**) Photographs of CG-2 when it has reached its half-life.

Under the condition of low water quantity, both agents formed stable foam with liquid CO_2 and supercritical CO_2, but the foam state varied greatly. The phenomenon of lower foaming height and a longer half-life is commonly observed in liquid CO_2, and in supercritical CO_2, the foaming height is higher, but the half-life is significantly shorter. Because of the higher density of liquid CO_2, the phenomenon of molecular thermal movement is weaker, and the volume of foam formed with water is significantly smaller, which is more stable under the maintenance of high pressure. Supercritical CO_2 is less dense. The phenomenon of molecular thermal movement is weaker, which leads to a higher foaming height, and at the same time, because of the presence of high temperature, the stability of the foam is significantly weakened and the half-life decreases dramatically. Through the experiments, it was found that the foaming effect of CG-1 with supercritical CO_2 was better, and the foaming effect of CG-2 with liquid CO_2 was better.

According to the experimental results, the foaming volume of the blowing agent and liquid CO_2 is low. Therefore, after mixing thoroughly by stirring, the foam occupies only a tiny part of the volume in the piston. The upper part of the foam is a mixture of CO_2 and the blowing agent, and the foam will not be injected into the rock center during the replacement, which ensures a pure gas injection environment (as shown in Figure 7).

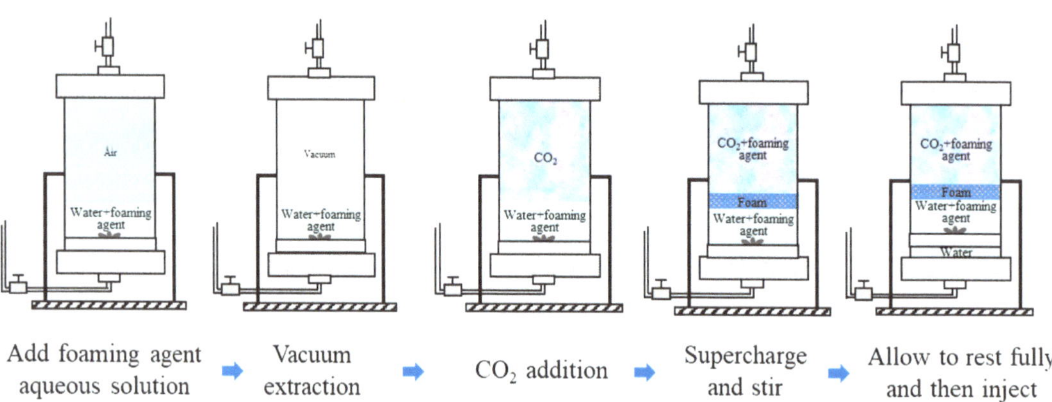

Figure 7. Configuration of CO_2 + blowing agent mixtures.

3.1.2. Comparison of the Foaming Capacity of the Blowing Agent and Supercritical CO_2 at Different Water Contents

Since supercritical CO_2 was used to foam with water in the replacement experiments, focusing on the foaming performance of the agent with supercritical CO_2 was necessary. The temperature and pressure of foaming remained unchanged at 70 °C and 20 MPa, and the concentration was 1.6%. Based on 10% water content (25 mL), two test points of 5% (12.5 mL) and 20% (50 mL) were added to examine the foaming performance of each agent.

Table 2 presents the foaming data of agents and supercritical CO_2 under different water content conditions. It can be seen that the rise in water content plays a prominent role in promoting the foaming effect. Under the condition of 5% water content, the foaming height and half-life of CG-1 were shortened (0.4 cm/50 min), and CG-2 could not even foam. At 20% water content, the foaming height and half-life of CG-1 were 5.5 cm and 250 min, respectively, with a half-life 1.5 times that of 10% water, and the foaming height and half-life of CG-2 were 4.6 cm and 70 min, respectively, with a half-life 4.67 times that of 10% water. As shown in the foaming data, CG-1 also had a specific foaming ability in the case of low water content, and the scope of application was more comprehensive; CG-2 only had a better foaming effect in high water content conditions.

Table 2. Foaming effect of pharmaceuticals with supercritical CO_2 under different water content conditions.

Sample Number	CO_2 Phase	5% Water Content	10% Water Content	20% Water Content
CG-1	Supercritical	0.4 cm/50 min	4.5 cm/100 min	5.5 cm/250 min
CG-2	Supercritical	Non-foamable	3.7 cm/15 min	4.6 cm/70 min

3.1.3. Comparison of the Foaming Capacity of the Blowing Agent and Supercritical CO_2 at Different Water Contents

Oil and water coexist in the rock pore throat, so the foaming effect of the agent in the case of oil content must be examined. Water–oil ratios of 100:3 (0.75 mL), 100:10 (2.5 mL), and 100:20 (5 mL) were set to investigate the effect of oil content on the foaming performance of the supercritical CO_2 foaming system. The temperature and pressure were also 70 °C and 20 MPa, the concentration was 1.6%, and the water content was 10% (25 mL) unchanged. In order to measure the effect of oil content on foaming performance more accurately, the "foam factor" was introduced as a non-factorized quantity, where the higher the foam factor, the better the foaming performance (foam factor = 0.75 × foaming height × half-life).

As shown in Table 3, with the increase in oil content, the foam factor showed a tendency to increase and then decrease. In the case of low oil content, the oil on the two foaming agents showed a prominent promotion of foaming. The foaming effect of CG-2 was pronounced after adding oil, where the foam factor became 6–10 times the case of no oil, and the foaming performance increased significantly. The foam factor of CG-1 after refueling became 0.75–2.15 times the original, and there was a phenomenon of weakening the foaming effect at high oil content. The table shows that although the effect of oil on CG-1 is relatively tiny, overall, the foaming effect of CG-1 after refueling is still better than that of CG-2.

Table 3. Foaming effect of pharmaceuticals with supercritical CO_2 under different oil content conditions.

Sample Number	100:0	100:3	100:10	100:20
CG-1	4.5 cm/100 min 337.5	4.5 cm/120 min 405	4.8 cm/200 min 720	4.5 cm/75 min 253.125
CG-2	3.7 cm/15 min 41.625	4 cm/90 min 270	4.3 cm/140 min 451.5	4.6 cm/100 min 345

3.1.4. Foaming Agent Screening

According to the results of the above studies, it can be found that the foaming effect of CG-2 with liquid CO_2 is better than that of CG-1, but the foaming effect of CG-1 with supercritical CO_2 is better than that of CG-2. CG-1 can be foamed under low water content conditions, and its foaming effect is better than that of CG-2 in various water content conditions. Although the promotion effect of oil on the foaming performance of CG-2 is better than that of CG-1, the oil content of CG-1 is better than that of CG-2. Although the promotion effect of oil on the foaming performance of CG-2 is better than that of CG-1, the foaming effect of CG-1 in oil content is better than that of CG-2 on the whole. Based on the above experimental results, it is assumed that CG-1 has a better foaming effect with supercritical CO_2 and performs better under low water conditions. At the same time, CG-1 still maintains an excellent foaming effect under oil-containing conditions. Therefore, CG-1 was selected to investigate the effect of the use of the agent in fractured rock cores.

3.2. Effectiveness of CO_2-Soluble Blowing Agent in Fractured Rock Cores

Because of the low permeability of the core matrix (about 0.25 mD), it is not possible to use water drive development. Therefore, the CO_2-soluble foaming agent was dissolved in CO_2 and injected into the core, which foamed with the oil and water on the fracture surface to form foam and then sealed the core, thus increasing the recovery. Cores with one fully open fracture were used for this study.

3.2.1. Influence of the Injection Rate on the Repellent Characteristics of a Fully Open Fracture Core (Blank Controlled Experiment)

The gas flow rate graphs of the core with one fully open fracture at different injection rates are shown in Figure 8. With the increase in the injection flow rate from 0.05 to 0.8 mL/min, the gas production rate in the late stage of replacement gradually increased from 421.36 to 1326.86 mL/(MPa·min).

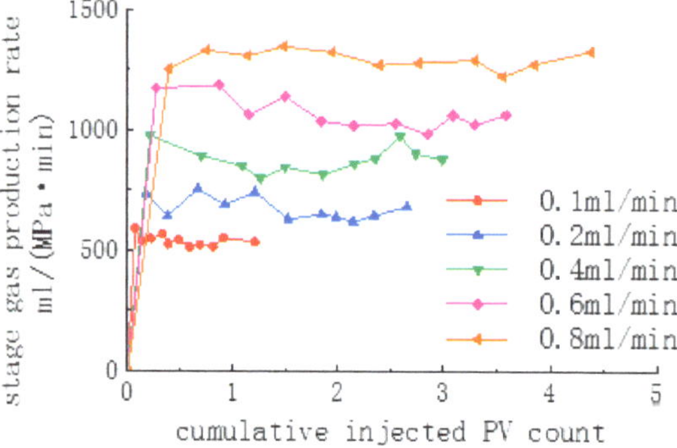

Figure 8. Gas flow rates at different injection rates.

The effect of different injection flow rates on the degree of replacement from a fully open fracture core is shown in Figure 9. The figure shows that as the injection flow rate rises, the replacement degree shows a tendency to first increase and then decrease. In the replacement process of a fractured core, oil is first transported from the matrix into the fracture and subsequently carried out of the core by the CO_2 + blowing agent. Too low a flow rate results in a low amount of CO_2 being injected, while at the same time, the differential pressure is low (as shown in Figure 10, the higher the flow rate, the higher the differential pressure). The amount of diffusion into the fracture will be even lower, preventing the effective movement of crude oil from the matrix. However, if the flow rate is too high, although the differential pressure rises, in that case, the flow rate of the gas is too fast, which is also unfavorable to the diffusion of CO_2 into the matrix, leading to a decrease in the degree of recovery.

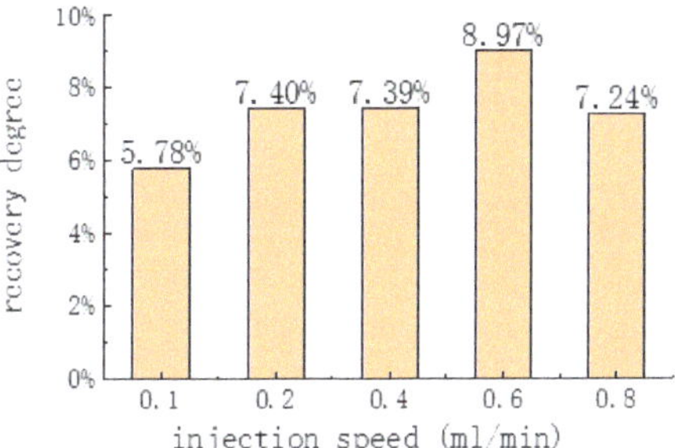

Figure 9. Influence of different injection velocities on the fracture core recovery degree.

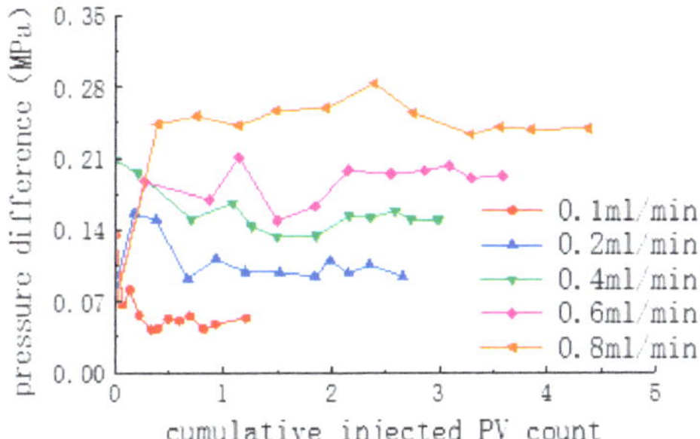

Figure 10. Displacement pressure difference at different injection flow rates.

3.2.2. Effect of Injection Velocity on Displacement Characteristics of the Foaming Agent

Figure 11 shows the stage gas production rates at different injection rates for CO_2 replacement of a core with one fully open fracture with both CG-1 concentrations of 1.11%. The graphs show the gas production rates when the injection rate increases from 0.1 mL/min to 0.8 mL/min. For the pure CO_2 drive stage, the injection rate was 0.4 mL/min in all cases, and the stage gas production rate range was roughly 850–1200 mL/(MPa-min) because of the difference in fracture morphology. The gas production rate was suppressed immediately after injection of the blowing agent (including the experimental group of 0.4–0.8 mL/min), showing a gradual decrease in the gas production rate, which quickly returned to the previous state after reinjection of CO_2.

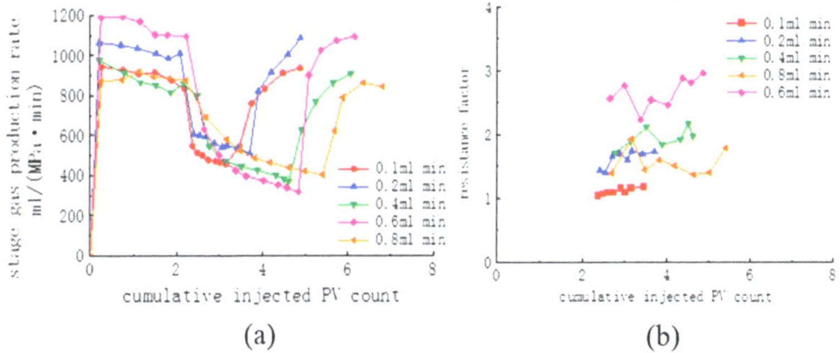

Figure 11. Stage gas production rates (**a**) and the resistance factor (**b**) at different injection rates.

Since the gas production rate is affected by both the injection rate of the system and the agent, the injection rate of the CO_2 + foaming agent is not the same as that of pure CO_2. Therefore, the resistance factor was used to examine the blocking performance of the agent by calculating the ratio of the injection pressure difference at the late stage of the replacement at the same injection rate (the pressure was unstable in the early stage of the replacement, and it took time for the foam to form). The resistance factor for foam formation at each injection rate is shown in Table 4.

Table 4. Resistance factor for blowing agent formation at each injection rate.

injection rate mL/min	0.1	0.2	0.4	0.6	0.8
resistance factor	1.17	1.73	2.20	2.95	1.80

Table 4 and Figure 11b show that the resistance factor increases and then decreases as the injection rate increases. The resistance factor was higher at 0.4 and 0.6 mL/min measuring 2.20 and 2.95, respectively. In the static foaming evaluation, foaming was carried out by stirring, which stopped after the foaming was completed. However, in the replacement process, without any stirring device, it can only rely on the flow of gas to provide power, in turn forming foam on the fracture surface, which is both the foam's storage space and the gas's main flow channel. Therefore, the increased gas flow both promotes foam formation and leads to foam destruction, so a decreasing drag factor at a high flow rate occurs.

3.2.3. Influence of the Injection Concentration on the Effect of Expulsion

According to the results in Table 4, the injection flow rates of the agents, 0.4 mL/min and 0.6 mL/min, were preferably selected. The blowing agent concentrations were set to 0.56%, 1.11%, 1.67%, 2.28%, and 2.80% to investigate the influence of the blowing agent concentration on the effect of the replacement (0.4 mL/min in all cases of pure CO_2 drive). Figure 12 shows a comparison of the stage gas production rate, resistance factor, and sealing rate at different injection concentrations when the agent injection rate is 0.4 mL/min (at this time, the injection rate of the CO_2 + blowing agent is the same as that of the pure CO_2 replacement, and the existence of the blocking rate is meaningful). It was found that at an injection rate of 0.4 mL/min, CG-1 had a superior sealing performance at all concentrations, and the sealing performance gradually increased with the increase in injection concentration. At the injection concentration of 2.8%, the resistance factor in the late replacement stage was 6.1, the blocking rate was 80.72%, and the stage gas production rate was 274 mL/(MPa·min).

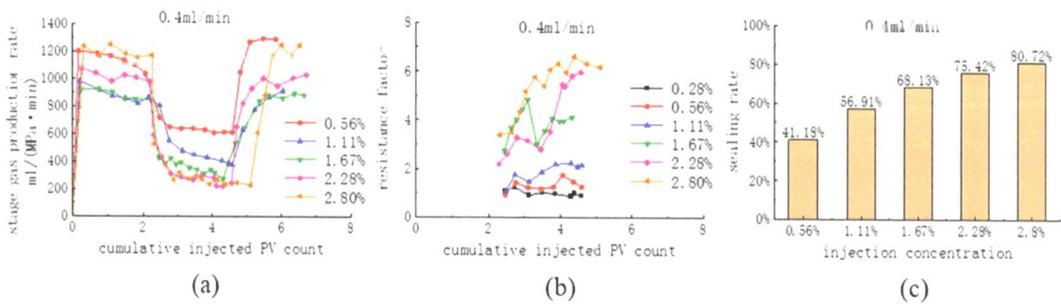

Figure 12. Comparison of (**a**) the stage gas production rate, (**b**) the drag factor, and (**c**) the blocking rate at different injection concentrations (0.4 mL/min).

Figure 13 shows a comparison of the stage gas production rate, resistance factor, and sealing rate at different injection concentrations at the agent injection rate of 0.6 mL/min.

At an injection concentration of 2.8%, the resistance factor in the late replacement stage was 6.3, the sealing rate was 83.775%, and the stage gas production rate was 200 mL/(MPa·min). After increasing the injection rate from 0.4 mL/min to 0.6 mL/min, there was an increase in the sealing performance at all concentrations, and in terms of the sealing rate, the increase in the sealing effect by increasing the injection rate was not pronounced. However, from the early injection stage of the system, the onset time of the agent was shortened, the sealing was more rapid, the rate of gas production in the stage decreased more rapidly, and the fluctuation in the resistance factor was significantly weakened. It can be considered that the foam is more stable at 0.6 mL/min.

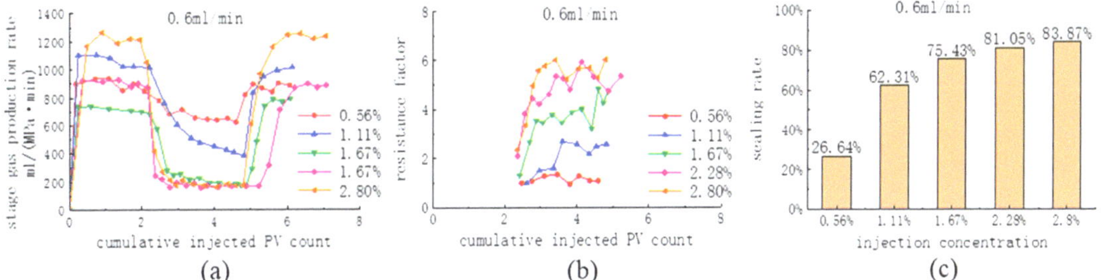

Figure 13. Comparison of (**a**) the stage gas production rate, (**b**) the drag factor, and (**c**) the blocking rate at different injection concentrations (0.6 mL/min).

A comparison of CG-1 sealing rate at different injection rates and concentrations is shown in Figure 14a. At a CG-1 concentration of 0.56%, the injection flow rate of 0.4 mL/min has a better sealing effect; when the CG-1 concentration is more significant than 1.11%, it has a better sealing effect at an injection flow rate of 0.6 mL/min. Through static experiments, it was found that when the concentration is low, the foaming volume and stability are reduced. The injection rate is high, which promotes foaming, but it currently has a stronger destructive ability, leading to a decrease in the sealing rate, so it is suitable to use a low injection flow rate. The foam is stable at high concentrations and can resist higher injection flow rates, while a high flow rate promotes system foaming and strengthens the sealing performance.

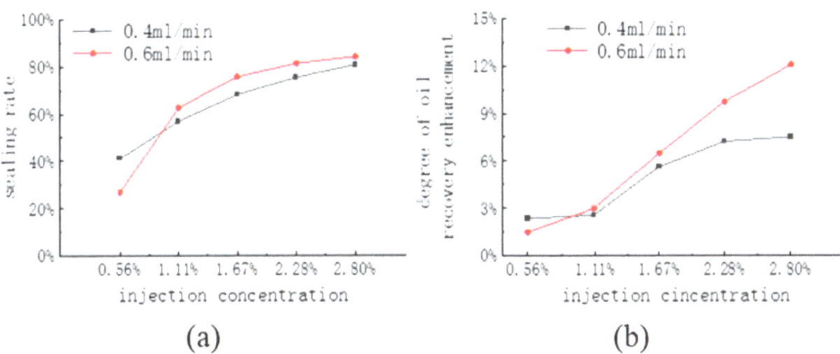

Figure 14. Comparison of CG-1 (**a**) the sealing rate and (**b**) enhanced recovery at different injection rates and concentrations.

A comparison of the degree of enhanced recovery of CG-1 at different injection rates and concentrations is shown in Figure 14b. Although a fully fractured core is relatively controllable during the fracturing process, ensuring complete consistency between cores is challenging, and the degree of recovery varies significantly among different cores. Therefore, the sealing performance was evaluated using the recovery enhancement after the agent was injected into the core. The enhancement of recovery of CG-1 was 1–3% when the concentration was less than 1.67%, but the enhancement of recovery was 6–12.5% when the concentration ranged from 1.67% to 2.80%, which had a significant effect. The sealing performance of CG-1 at a high injection concentration and a high flow rate was excellent, with a sealing rate above 80%, which significantly increased the proportion of CO_2 diffusion into the matrix and, therefore, improved the degree of recovery significantly. In the case of high flow rate injection, the amount of CO_2 injected into the core for the same length of time was significantly higher, increasing the injection differential pressure and, at the same time, increasing the proportion of CO_2 diffusing into the core matrix. This led to a significant increase in the degree of increased recovery at the higher flow rate for two injection flow rates that are close to each other in terms of sealing rate.

4. Conclusions

This paper focuses on the foaming performance of CO_2-soluble blowing agents under different conditions and their effectiveness in sealing the fractured core and enhancing recovery. Based on a comparison of the foaming effect of foaming agents in different phases, water contents, and oil contents, CG-1 is preferred for evaluating the replacement effect, which has a good use effect. The specific conclusions are as follows:

1. Through static experimental evaluation, it is found that CG-1 can produce stable foam under low water content and various oil content conditions. The foaming conditions are more relaxed, with a broader range of use, which is suitable for use in reservoirs with variable conditions and can be applied to the subsequent evaluation of the replacement experiments.
2. Because of the low foaming height of both blowing agents with liquid CO_2, a large amount of CO_2 and blowing agent mixture exists in the upper part of the space, which ensures that there is no participation of any foam or water in the injection process and provides a method for the configuration and injection of the agent.
3. In the dynamic replacement experiment, if the injection rate is too low, the foaming agent cannot form stable foam with the water and oil on the fracture surface. However, if the injection rate is too high, the airflow will simultaneously destroy the foam that has already been formed, so that the sealing rate will be reduced, which is unfavorable to development. At 0.6 mL/min and 2.80% injection, the sealing rate can reach 83.7%, and the improvement in the extraction degree is 12.02%, which has a good use effect.

Author Contributions: Conceptualization, M.L., K.S. and H.F.; methodology, M.L. and K.S.; software, M.L. and L.W.; validation, M.L., L.W. and J.Z.; formal analysis, M.L. and J.Z.; investigation, M.L. and L.W.; resources, H.F.; data curation, M.L., H.F. and J.Z.; writing—original draft preparation, M.L., J.Z. and L.W.; writing—review and editing, H.F. and J.Z.; visualization, L.W.; supervision H.F.; project administration, K.S.; funding acquisition, M.L. and K.S. All authors have read and agreed to the published version of this manuscript

Funding: This research was funded by [the National Natural Science Foundation of China] grant number [5183000475] and [Experimental study on CO_2 flooding rule and control technology in typical offshore oil fields] grant number [CCL2022RCPS0400KSN].

Institutional Review Board Statement: Not applicable.

Data Availability Statement: The data presented in this study are available on request from the first author. (The data is not publicly available due to confidentiality policies).

Acknowledgments: Thanks to Zhihui Yang and Zhanjun Sun of Hebei Ruiniu Company for their experimental technical guidance. Thanks to Jian Ding and Hai Liu for the experimental equipment and Baodong Gu and Yinghua Zhang for the experimental core and fracturing equipment.

Conflicts of Interest: We would like to submit the enclosed manuscript titled "Application of CO_2-soluble polymer-based blowing agent to improve supercritical CO_2 replacement in low-permeability fractured reservoirs", for publication in *Polymers*. No conflicts of interest exists regarding the submission of this manuscript, which has been approved by all authors for publication. I would like

Polymers **2024**, *16*, 2191

to declare on behalf of my coauthors that the research described is original, has not been published previously, and is not under consideration for publication elsewhere, in whole or in part. All listed authors approved the enclosed manuscript.

References

1. Cao, A.; Li, Z.; Zheng, L.; Bai, H.; Zhu, D.; Li, B. Nuclear Magnetic Resonance Study of CO_2 Flooding in Tight Oil Reservoirs: Effects of Matrix Permeability and Fracture. *Geoenergy Sci. Eng.* **2023**, *225*, 211692. [CrossRef]
2. Wang, Y.; Song, X.; Tian, C.; Shi, C.; Li, J.; Hui, G.; Hou, J.; Gao, C.; Wang, X.; Liu, P. Dynamic Fractures Are an Emerging New Development Geological Attribute in Water-Flooding Development of Ultra-Low Permeability Reservoirs. *Pet. Explor. Dev.* **2015**, *42*, 247–253. [CrossRef]
3. Zhao, K.; Jia, C.; Li, Z.; Du, X.; Wang, Y.; Li, J.; Yao, Z.; Yao, J. Recent Advances and Future Perspectives in Carbon Capture, Transportation, Utilization, and Storage (CCTUS) Technologies: A Comprehensive Review. *Fuel* **2023**, *351*, 128913. [CrossRef]
4. Bai, H.; Zhang, Q.; Li, Z.; Li, B.; Zhu, D.; Zhang, L.; Lv, G. Effect of Fracture on Production Characteristics and Oil Distribution during CO_2 Huff-n-Puff under Tight and Low-Permeability Conditions. *Fuel* **2019**, *246*, 117–125. [CrossRef]
5. Yang, M.; Huang, S.; Zhao, F.; Sun, H.; Chen, X. Experimental Investigation of CO_2 Huff-n-Puff in Tight Oil Reservoirs: Effects of the Fracture on the Dynamic Transport Characteristics Based on the Nuclear Magnetic Resonance and Fractal Theory. *Energy* **2024**, *294*, 130781. [CrossRef]
6. Wan, T.; Zhang, J.; Dong, Y. The Impact of Fracture Network on CO_2 Storage in Shale Oil Reservoirs. *Geoenergy Sci. Eng.* **2023**, *231*, 212322. [CrossRef]
7. Dang, F.; Li, S.; Dong, H.; Wang, Z.; Zhu, J. Experimental Study on the Oil Recovery Characteristics of CO_2 Energetic Fracturing in Ultralow-Permeability Sandstone Reservoirs Utilizing Nuclear Magnetic Resonance. *Fuel* **2024**, *366*, 131370. [CrossRef]
8. Zhang, J.; Zhang, H.X.; Ma, L.Y.; Liu, Y.; Zhang, L. Performance Evaluation and Mechanism with Different CO_2 Flooding Modes in Tight Oil Reservoir with Fractures. *J. Pet. Sci. Eng.* **2020**, *188*, 106950. [CrossRef]
9. Wei, B.; Zhong, M.; Gao, K.; Li, X.; Zhang, X.; Cao, J.; Lu, J. Oil Recovery and Compositional Change of CO_2 Huff-n-Puff and Continuous Injection Modes in a Variety of Dual-Permeability Tight Matrix-Fracture Models. *Fuel* **2020**, *276*, 117939. [CrossRef]
10. Chen, H.; Li, H.; Li, Z.; Li, S.; Wang, Y.; Wang, J.; Li, B. Effects of Matrix Permeability and Fracture on Production Characteristics and Residual Oil Distribution during Flue Gas Flooding in Low Permeability/Tight Reservoirs. *J. Pet. Sci. Eng.* **2020**, *195*, 107813. [CrossRef]
11. Milad, M.; Junin, R.; Sidek, A.; Imqam, A.; Alusta, G.A.; Augustine, A.; Abdulazeez, M.A. Experimental Investigation of Bypassed-Oil Recovery in Tight Reservoir Rock Using a Two-Step CO_2 Soaking Strategy: Effects of Fracture Geometry. *Upstream Oil Gas Technol.* **2023**, *11*, 100093. [CrossRef]
12. Wang, Y.-Z.; Cao, R.-Y.; Jia, Z.-H.; Wang, B.-Y.; Ma, M.; Cheng, L.-S. A Multi-Mechanism Numerical Simulation Model for CO_2-EOR and Storage in Fractured Shale Oil Reservoirs. *Pet. Sci.* **2024**, *21*, 1814–1828. [CrossRef]
13. Panja, P.; Deo, M. Unusual Behavior of Produced Gas Oil Ratio in Low Permeability Fractured Reservoirs. *J. Pet. Sci. Eng.* **2016**, *144*, 76–83. [CrossRef]
14. Zhao, Y.; Song, Y.; Wang, T.; Liu, Y.; Jiang, L.; Zhu, N.; Yang, W. Visualisation of Water Flooding and Subsequent Supercritical CO_2 Flooding in Fractured Porous Media with Permeability Heterogeneity Using MRI. *Energy Procedia* **2013**, *37*, 6942–6949. [CrossRef]
15. Wang, Y.; Shang, Q.; Zhou, L.; Jiao, Z. Utilizing Macroscopic Areal Permeability Heterogeneity to Enhance the Effect of CO_2 Flooding in Tight Sandstone Reservoirs in the Ordos Basin. *J. Pet. Sci. Eng.* **2021**, *196*, 107633. [CrossRef]
16. Cong, S.; Li, J.; Liu, W.; Shi, Y.; Li, Y.; Zheng, K.; Luo, X.; Luo, W. EOR Mechanism of Fracture Oil Displacement Agent for Ultra-Low Permeability Reservoir. *Energy Rep.* **2023**, *9*, 4893–4904. [CrossRef]
17. Li, N.; Zhu, S.; Li, Y.; Zhao, J.; Long, B.; Chen, F.; Wang, E.; Feng, W.; Hu, Y.; Wang, S.; et al. Fracturing-Flooding Technology for Low Permeability Reservoirs: A Review. *Petroleum* **2024**, *10*, 202–215. [CrossRef]
18. Li, Y.-K.; Hou, J.-R.; Wu, W.-P.; Qu, M.; Liang, T.; Zhong, W.-X.; Wen, Y.-C.; Sun, H.-T.; Pan, Y.-N. A Novel Profile Modification HPF-Co Gel Satisfied with Fractured Low Permeability Reservoirs in High Temperature and High Salinity. *Pet. Sci.* **2024**, *21*, 683–693. [CrossRef]
19. Li, L.; Xiuting, H.; Lin, M. Experimental Research on Profile Control for Oil Displacement by Functional Polymer in Low Permeability Fractured Reservoir. *Phys. Procedia* **2012**, *25*, 1292–1300. [CrossRef]
20. Zhang, C.; Wang, P.; Song, G.; Qu, G.; Liu, J. Optimization and Evaluation of Binary Composite Foam System with Low Interfacial Tension in Low Permeability Fractured Reservoir with High Salinity. *J. Pet. Sci. Eng.* **2018**, *160*, 247–257. [CrossRef]
21. Du, D.; Chen, B.; Pu, W.; Zhou, X.; Liu, R.; Jin, F. CO_2-Responsive Gel Particles and Wormlike Micelles Coupling System for Controlling CO_2 Breakthrough in Ultra-Low Permeability Reservoirs. *Colloids Surf. A Physicochem. Eng. Asp.* **2022**, *650*, 129546. [CrossRef]
22. Jiang, H.-Z.; Yang, H.-B.; Pan, R.-S.; Ren, Z.-Y.; Kang, W.-L.; Zhang, J.-Y.; Pan, S.-L.; Sarsenbekuly, B. Performance and Enhanced Oil Recovery Efficiency of an Acid-Resistant Polymer Microspheres of Anti-CO_2 Channeling in Low-Permeability Reservoirs. *Pet. Sci.* **2024**, S1995822624000384. [CrossRef]
23. Hao, H.; Hou, J.; Zhao, F.; Song, Z.; Hou, L.; Wang, Z. Gas Channeling Control during CO 2 Immiscible Flooding in 3D Radial Flow Model with Complex Fractures and Heterogeneity. *J. Pet. Sci. Eng.* **2016**, *146*, 890–901. [CrossRef]

24. Zhao, F.; Wang, P.; Huang, S.; Hao, H.; Zhang, M.; Lu, G. Performance and Applicable Limits of Multi-Stage Gas Channeling Control System for CO_2 Flooding in Ultra-Low Permeability Reservoirs. *J. Pet. Sci. Eng.* **2020**, *192*, 107336. [CrossRef]

25. Li, D.; Saraji, S.; Jiao, Z.; Zhang, Y. An Experimental Study of CO_2 Injection Strategies for Enhanced Oil Recovery and Geological Sequestration in a Fractured Tight Sandstone Reservoir. *Geoenergy Sci. Eng.* **2023**, *230*, 212166. [CrossRef]

26. Fernø, M.A.; Steinsbø, M.; Eide, Ø.; Ahmed, A.; Ahmed, K.; Graue, A. Parametric Study of Oil Recovery during CO_2 Injections in Fractured Chalk: Influence of Fracture Permeability, Diffusion Length and Water Saturation. *J. Nat. Gas Sci. Eng.* **2015**, *27*, 1063–1073. [CrossRef]

27. Li, X.; Pu, C.; Wei, H.; Huang, F.; Bai, Y.; Zhang, C. Enhanced Oil Recovery in Fractured Low-Permeability Reservoirs by a Novel Gel System Prepared by Sustained-Release Crosslinker and Water-Soluble Thixotropic Polymer. *Geoenergy Sci. Eng.* **2023**, *222*, 211424. [CrossRef]

28. Zhu, W.; Ma, Q.; Han, H. Theoretical Study on Profile Control of a Nano-Microparticle Dispersion System Based on Fracture–Matrix Dual Media by a Low-Permeability Reservoir. *Energy Rep.* **2021**, *7*, 1488–1500. [CrossRef]

29. Du, D.; Zou, B.; Pu, W.; Wei, X.; Liu, R. Injectivity and Plugging Characteristics of CO_2-Responsive Gel Particles for Enhanced Oil Recovery in Fractured Ultra-Low Permeability Reservoirs. *J. Pet. Sci. Eng.* **2022**, *214*, 110591. [CrossRef]

30. Youssif, M.I.; Sharma, K.V.; Shoukry, A.E.; Goual, L.; Piri, M. Methane Foam Performance Evaluation in Fractured Oil-Wet Carbonate Systems at Elevated Pressure and Temperature Conditions. *J. Environ. Chem. Eng.* **2024**, *12*, 112444. [CrossRef]

31. Wang, F.; Zhang, Y.; Yu, Z.; Chen, J. Simulation Research on Jamin Effect and Oil Displacement Mechanism of CO_2 Foam under Microscale. *Arab. J. Chem.* **2023**, *16*, 105123. [CrossRef]

32. Xu, Z.; Li, Z.; Liu, Z.; Li, B.; Zhang, Q.; Zheng, L.; Song, Y.; Husein, M.M. Characteristics of CO_2 Foam Plugging and Migration: Implications for Geological Carbon Storage and Utilization in Fractured Reservoirs. *Sep. Purif. Technol.* **2022**, *294*, 121190. [CrossRef]

33. Mansha, M.; Ali, S.; Alsakkaf, M.; Karadkar, P.B.; Harbi, B.G.; Yamani, Z.H.; Khan, S.A. Advancements in Nanoparticle-Based Stabilization of CO_2 Foam: Current Trends, Challenges, and Future Prospects. *J. Mol. Liq.* **2023**, *391*, 123364. [CrossRef]

34. Issakhov, M.; Shakeel, M.; Pourafshary, P.; Aidarova, S.; Sharipova, A. Hybrid Surfactant-Nanoparticles Assisted CO_2 Foam Flooding for Improved Foam Stability: A Review of Principles and Applications. *Pet. Res.* **2022**, *7*, 186–203. [CrossRef]

35. Solbakken, J.S.; Aarra, M.G. CO_2 Mobility Control Improvement Using N_2-Foam at High Pressure and High Temperature Conditions. *Int. J. Greenh. Gas Control* **2021**, *109*, 103392. [CrossRef]

36. Monjezi, K.; Mohammadi, M.; Khaz'ali, A.R. Stabilizing CO_2 Foams Using APTES Surface-Modified Nanosilica: Foamability, Foaminess, Foam Stability, and Transport in Oil-Wet Fractured Porous Media. *J. Mol. Liq.* **2020**, *311*, 113043. [CrossRef]

37. Wu, Q.; Ding, L.; Zhao, L.; Alhashboul, A.A.; Almajid, M.M.; Patil, P.; Zhao, W.; Fan, Z. CO_2 Soluble Surfactants for Carbon Storage in Carbonate Saline Aquifers with Achievable Injectivity: Implications from the Continuous CO_2 Injection Study. *Energy* **2024**, *290*, 130064. [CrossRef]

38. Ren, G. Assess the Potentials of CO_2 Soluble Surfactant When Applying Supercritical CO_2 Foam. Part I: Effects of Dual Phase Partition. *Fuel* **2020**, *277*, 118086. [CrossRef]

39. Føyen, T.; Alcorn, Z.P.; Fernø, M.A.; Barrabino, A.; Holt, T. CO_2 Mobility Reduction Using Foam Stabilized by CO_2- and Water-Soluble Surfactants. *J. Pet. Sci. Eng.* **2021**, *196*, 107651. [CrossRef]

40. Barrabino, A.; Holt, T.; Lindeberg, E. Partitioning of Non-Ionic Surfactants between CO_2 and Brine. *J. Pet. Sci. Eng.* **2020**, *190*, 107106. [CrossRef]

41. Hu, D.; Gu, Y.; Liu, T.; Zhao, L. Microcellular Foaming of Polysulfones in Supercritical CO_2 and the Effect of Co-Blowing Agent. *J. Supercrit. Fluids* **2018**, *140*, 21–31. [CrossRef]

42. Zhang, C.; Wu, P.; Li, Z.; Liu, T.; Zhao, L.; Hu, D. Ethanol Enhanced Anionic Surfactant Solubility in CO_2 and CO_2 Foam Stability: MD Simulation and Experimental Investigations. *Fuel* **2020**, *267*, 117162. [CrossRef]

43. Zeng, Y.; Farajzadeh, R.; Biswal, S.L.; Hirasaki, G.J. A 2-D Simulation Study on CO_2 Soluble Surfactant for Foam Enhanced Oil Recovery. *J. Ind. Eng. Chem.* **2019**, *72*, 133–140. [CrossRef]

44. Farajzadeh, R.; Andrianov, A.; Krastev, R.; Hirasaki, G.J.; Rossen, W.R. Foam–Oil Interaction in Porous Media: Implications for Foam Assisted Enhanced Oil Recovery. *Adv. Colloid Interface Sci.* **2012**, *183–184*, 1–13. [CrossRef]

45. Ren, G.; Yu, W. Numerical Investigations of Key Aspects Influencing CO_2 Foam Performance in Fractured Carbonate System Using CO_2 Soluble Surfactants. *J. CO2 Util.* **2019**, *33*, 96–113. [CrossRef]

 *polymers*

Article

Experimental Study on the Application of Polymer Agents in Offshore Oil Fields: Optimization Design for Enhanced Oil Recovery

Xianjie Li [1,2], Jian Zhang [1,2], Yaqian Zhang [3,4], Cuo Guan [1,2], Zheyu Liu [3,4,*], Ke Hu [1,2], Ruokun Xian [1,2] and Yiqiang Li [3,4]

[1] State Key Laboratory of Offshore Oil and Gas Exploitation, Beijing 100028, China; lixj8@cnooc.com.cn (X.L.); zhangjian@cnooc.com.cn (J.Z.); guancuo@cnooc.com.cn (C.G.); huke@cnooc.com.cn (K.H.); xianrk@cnooc.com.cn (R.X.)
[2] CNOOC Research Institute Co., Ltd., Beijing 100028, China
[3] State Key Laboratory of Petroleum Resources and Engineering, China University of Petroleum (Beijing), Beijing 102249, China; yaqianz_cup@163.com (Y.Z.); yiqiangli@cup.edu.cn (Y.L.)
[4] College of Petroleum Engineering, China University of Petroleum (Beijing), Beijing 102249, China
* Correspondence: zheyu.liu@cup.edu.cn

Academic Editor: Bob Howell

Received: 29 November 2024
Revised: 15 January 2025
Accepted: 15 January 2025
Published: 20 January 2025

Citation: Li, X.; Zhang, J.; Zhang, Y.; Guan, C.; Liu, Z.; Hu, K.; Xian, R.; Li, Y. Experimental Study on the Application of Polymer Agents in Offshore Oil Fields: Optimization Design for Enhanced Oil Recovery. *Polymers* **2025**, *17*, 244. https://doi.org/10.3390/polym17020244

Abstract: The Bohai oilfield is characterized by severe heterogeneity and high average permeability, leading to a low water flooding recovery efficiency. Polymer flooding only works for a certain heterogeneous reservoir. Therefore, supplementary technologies for further enlarging the swept volume are still necessary. Based on the concept of discontinuous chemical flooding with multi slugs, three chemical systems, which were polymer gel (PG), hydrophobically associating polymer (polymer A), and conventional polymer (polymer B), were selected as the profile control and displacing agents. The optimization design of the discontinuous chemical flooding was investigated by core flooding experiments and displacement equilibrium degree calculation. The gel, polymer A, and polymer B were classified into three levels based on their profile control performance. The degree of displacement equilibrium was defined by considering the sweep conditions and oil displacement efficiency of each layer. The effectiveness of displacement equilibrium degree was validated through a three-core parallel displacement experiment. Additionally, the parallel core displacement experiment optimized the slug size, combination method, and shift timing of chemicals. Finally, a five-core parallel displacement experiment verified the enhanced oil recovery (EOR) performance of discontinuous chemical flooding. The results show that the displacement equilibrium curve exhibited a stepwise change. The efficiency of discontinuous chemical flooding became more significant with the number of layers increasing and heterogeneity intensifying. Under the combination of permeability of 5000/2000/500 mD, the optimal chemical dosage for the chemical discontinuous flooding was a 0.7 pore volume (PV). The optimal combination pattern was the alternation injection in the form of "medium-strong-weak-strong-weak", achieving a displacement equilibrium degree of 82.3%. The optimal shift timing of chemicals occurred at a water cut of 70%, yielding a displacement equilibrium degree of 87.7%. The five-core parallel displacement experiment demonstrated that discontinuous chemical flooding could get a higher incremental oil recovery of 24.5% compared to continuous chemical flooding, which presented a significantly enhanced oil recovery potential.

Keywords: strong heterogeneity; discontinuous chemical flooding; polymer slug combination; displacement equilibrium

1. Introduction

Polymers are widely applied in various fields such as construction, healthcare, and energy [1–3]. One of their most extensive applications is in enhancing oil recovery through chemical flooding processes [4]. Currently, polymers in common use are classified into two types: synthetic polymers and biopolymers [3–5]. Synthetic polymers are artificially synthesized by the polymerization of acrylamide monomers and are often modified based on the requirement of the reservoir [6,7]. Hydrolyzed polyacrylamide (HPAM) is the most common and widely used synthetic polymer both at the field and experimental level due to its cost-effective nature [6,8]. Most synthetic polymers are PAM derivatives [4]. Polymer flooding can significantly increase crude oil recovery factors and is widely used among existing chemical flooding techniques, with broad application prospects [9,10]. The mechanisms for enhancing recovery primarily involve increasing the viscosity of the aqueous phase and reducing the aqueous phase permeability, thereby improving the oil–water mobility ratio, enhancing the vertical water absorption profile and increasing the areal sweep efficiency [8–10]. Additionally, at the microscopic scale, due to the polymer's (i) pulling effect, (ii) stripping effect, (iii) oil thread phenomenon, and (iv) shear thickening effect, polymers can mobilize residual oil trapped by capillary forces and rock structures, thus improving microscopic oil displacement efficiency [9–11]. As a mature chemical flooding technology for enhanced oil recovery, polymer flooding is widely applied in onshore oilfields such as Daqing, Dagang, Yumen, and Shengli, and has achieved remarkable results in improving oil recovery factors [8,9,12].

Offshore oilfields typically comprise loose sandstone with high permeability, strong reservoir heterogeneity, and high oil viscosity. The large well spacing, multilayer commingled development, and strong injection and production will result in the severe channeling of injected fluids, ultimately leading to a low water flooding efficiency. There are still a lot of residual oil remains after water flooding in the reservoirs [13–15]. Polymer flooding can further expand the sweep volume based on water flooding, and is currently an effective technique for enhancing oil recovery (EOR) in offshore oilfields. However, the late stage of polymer flooding is inevitably accompanied by profile reversal, reducing the reservoir control capacity and leaving approximately 50% of the more dispersed oil in the formation [7,16,17]. Numerous laboratory and field tests have shown that polymer flooding can only increase oil recovery by about 10–12% over water flooding [18–21]. Furthermore, in the high-water-cut stage of offshore oilfields, reservoir heterogeneity is further exacerbated, and the profile control ability of polymers is limited. In the presence of dominant flow channels and large pores, the polymer cannot prevent channeling. Therefore, supplementary technologies are needed to address the limitations of polymer flooding in expanding the swept volume in offshore oilfields [22–24].

Foam flooding, gel particle flooding, and multiphase thermal fluid flooding can effectively alleviate fluid channeling and further enhance oil recovery in offshore oilfields [25–28]. However, these methods are more complex than polymer flooding. Based on the concepts of multi-slug injection of polymers [29] and stepwise injection [30], Zhang et al. proposed the theory and model of discontinuous chemical flooding for high-water-cut offshore oilfields [31,32]. This technology designs different concentrations and molecular weights of polymers, hydrophobically associating polymers, gel particles, and gels as slug combinations according to reservoir conditions, effectively controlling injection pressure and significantly increasing the swept volume while plugging channeling paths. The technology has achieved good field application results, but it is still in the laboratory research stage. Two main issues need further clarification: (1) how to evaluate the effect of discontinuous chemical flooding. Displacement equilibrium is commonly used to evaluate the development degree of heterogeneous reservoirs [20,31,32], but current indicators are difficult to monitor and obtain during laboratory and field applications, requiring a more straightforward and accurate method to evaluate the development effect of discontinuous chemical flooding. (2) Optimization design of slug combinations for discontinuous chemical flooding. The total injection volume of the polymer solution, the injection volume of each agent, the combination method, and the conversion timing all play a key role in the effect of discontinuous chemical flooding and require further laboratory research to provide a basis for field parameter optimization.

The evaluation method and parameter optimization design of discontinuous chemical flooding were systematically researched. First, the equilibrium displacement degree was defined based on the number of layers in heterogeneous reservoirs and the oil recovery factor in individual layers. To verify its feasibility and sensitivity, discontinuous flooding and single polymer flooding scenarios were designed. Using a large flat-panel model, the optimal slug size for discontinuous chemical flooding was determined. Three different intensities of profile control systems were selected, and a three-core parallel displacement experiment was conducted to optimize the combination method and injection timing of the agents. Subsequently, a five-cores parallel displacement experiment was used to verify the beneficial effects of discontinuous chemical flooding. The established evaluation method for discontinuous chemical flooding, applicable to typical offshore oilfields, effectively solve the problem of further enhancing oil recovery in the mid/late stages of polymer flooding in high-water-cut offshore oilfields.

2. Experimental Equipment and Process

2.1. Experimental Materials

The polymer displacement agents included gel, polymer A, and polymer B. The relative molecular masses of polymers A and B were 3×10^7 and 1×10^7, respectively, and both are odorless, white, solid powders. The gel was formulated from polymer B and crosslinking agents in a specific ratio, and its molecular structure is illustrated in Figure 1. The experimental cores included a square gel-encapsulated core, 4.5 cm $\times$ 4.5 cm $\times$ 30 cm, with gas permeability values of 500 mD, 2000 mD, 5000 mD, 7500 mD, and 10,000 mD (the last two permeabilities were used in a five-core parallel displacement experiment). Porosity was approximately 30% and oil saturation was around 65%. A flat gel-encapsulated core, 60 cm $\times$ 60 cm $\times$ 5 cm, with gas permeability values of 500 mD, 2000 mD, and 5000 mD was also used.

The oil used in the experiment was a simulated oil composed of field crude oil blended with aviation kerosene, with a viscosity of 45 mPa·s at 60 °C. The water used was simulated formation water with a salinity of 7153.35 mg/L, and the detailed ion composition is shown in Table 1.

Table 1. Formation water ion composition.

Ion	Na^+	K^+	Mg^{2+}	Ca^{2+}	Cl^-	SO_4^{2-}	Total
Concentration mg/L	2422.13	40.29	69.94	191.71	3971.52	17.67	
Ion	HCO_3^-	CO_3^{2-}	I^-	Br^-	B^-		7153.35
Concentration mg/L	392.57	47.52	0.36	4.47	3.39		

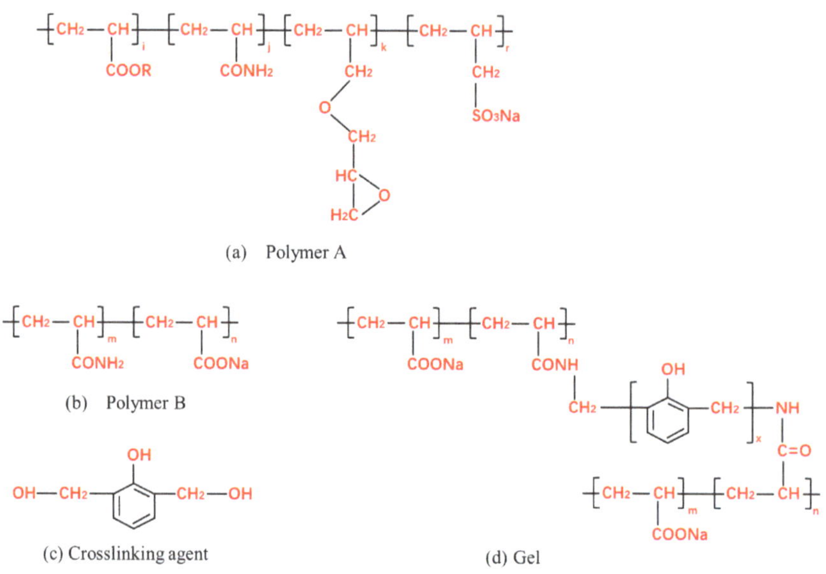

(a) Polymer A

(b) Polymer B

(c) Crosslinking agent

(d) Gel

Figure 1. Molecular structure.

2.2. Experimental Equipment

Polymer agent performance evaluation: a MARS III high-temperature and high-pressure rheometer (temperature range 10–300 °C) (HAAKE Company, Vreden, Germany), a TC-150SD viscometer (Brookfield Company, Mumbai, Maharashtra), an electric stirrer (Waring Company, Vicksburg, MI, USA).

Core oil displacement experiment: the polymer displacement system consisted of an injection system, a core model, a liquid measurement system, a pressure control system, and a data acquisition system. Major equipment included an ISCO constant speed pump (ISCO Company, Louisville, KY, USA), a thermostatic chamber (up to 100 °C) (Hai'an Petroleum Research Instrument Company, Haian, China), pressure sensors and a data acquisition box (Hai'an Petroleum Research Instrument Company).

2.3. Experimental Methods

2.3.1. Rheological Test

The rheological properties of the gel were tested using the HAAKE MARS III high-temperature and high-pressure rheometer. A prepared 12 mL sample was placed into a clean cylinder for stress and frequency scanning to obtain viscoelastic parameters.

2.3.2. Viscosity Test

The viscosity of the polymer displacement agents at 60 °C was measured using a TC-150SD Brookfield viscometer. For the gel agents, rotor No. 3 was used (test range 0–20,000 mPa·s) and, for polymer A and polymer B systems, rotor No. 1 was used (test range 0–200 mPa·s). Gel solutions were prepared by mixing the polymer and crosslinking agents, and the resultant mix was then aged at 60 °C for 72 h. The viscosity of the system was tested every 6 h to obtain a viscosity curve as a function of aging time. A 12 mL sample was placed in the clean, dry cylinder, and the rotor speed was set to 6 r/min. After stabilization, viscosity data were recorded.

2.3.3. Gel Strength Evaluation and Formula Optimization

First, the visual code method (Table 2) was used to select polymer concentrations that resulted in a gel strength between C and D and a suitable gelation time. Then, the optimal polymer-to-crosslinker ratio was determined using rheological parameters and viscosity variation methods [33].

Table 2. The gel strength code criteria [33].

Intensity Code	Gel Name	Corresponding Strength Description
A	Non-probabilistic gels	The system is no different from the polymer and is completely unbonded
B	High mobility gel	The viscosity of the system gradually increases and exceeds that of the polymer
C	Liquidity gel	Most of the gel can flow to the other end of the bottle
D	Medium flow gel	When flipping the glass bottle (<15% of the gel), the polymer cannot flow to the other end and often exists in the tongue type
E	Almost no flow of the gel	A small amount of the gel can flow slowly to the other end, and most of it is not liquid
F	High-shaped and immobile gel	The gel does not flow to the bottle mouth when turning around the glass bottle
G	Medium-shaped immobile gel	When flipped, it can only flow to the middle of the glass bottle
H	Slightly deformed immobile gel	Upon flipping, only the gel surface is deformed
I	Rigid gel	Upon flipping, the gel surface does not deform
J	Jolling gel	When shaking the glass bottle, you can feel the mechanical vibration like a sound fork

2.3.4. Definition of Displacement Equilibrium

The dynamic control degree of polymer flooding refers to the percentage of the pore volume in the oil layer that the polymer solution can affect under effective flow conditions, relative to the total pore volume of the oil layer. It reflects the extent of the reservoir sweep, which is related to reservoir pressure and dynamic resistance. The dynamic control degree of polymer flooding E is defined as:

$$E = \frac{\sum_{i=1}^{m} a_i b_i V(P)_i}{\sum_{j=1}^{n} V_i} \times 100\% \tag{1}$$

$$V(P) = \sum \left[\sum \left(S_{ji}(P) \cdot H_{ji}(P) \cdot \varphi \right) \right] \atop j=1, i=1 \tag{2}$$

where V is the pore volume of the oil layer that can be swept by the polymer solution and it is related to the effective driving pressure of the polymer. The terms a_i and b_i are dynamic control coefficients, related to dynamic resistance during polymer flooding. S_{ji} represents the sweep area of the polymer flood network for the i-th well group in the j-th oil layer, V_i is the total pore volume, H_{ji} is the connection thickness of the injection–production well group that the polymer molecules can reach, and ϕ is the porosity.

Increasing the dynamic control degree of the polymer reservoir can achieve a more uniform sweep. However, since the parameters of the control degree are difficult to determine directly, a simplified parameter based on experiments is needed to represent the sweep balance, known as the displacement equilibrium. The function of displacement equilibrium is to describe the sweep effect of the displacement system. Theoretically, the greater the sweep efficiency, the larger the sweep range and recovery factor. Displacement equilibrium λ is defined as:

$$\lambda = \xi \times \theta = \frac{n}{m} \times \frac{\eta_{\text{ave}}}{\eta_{\text{max}}} \times 100\% \tag{3}$$

where λ is the displacement equilibrium (as a percentage), ζ is the control coefficient, representing the ratio of the number of mobilized sub-layers n to the total number of sub-layers m, and θ is the homogeneity coefficient of mobilization, representing the ratio of the comprehensive recovery degree of each layer to the highest recovery degree of the layers.

2.3.5. Optimization Design of Discontinuous Chemical Flooding

Based on the characteristics of the reservoir's physical properties and thickness, artificial sandstone cores were designed according to the principle of similar permeability and thickness ratio, and displacement experiments were conducted. The experimental procedure was as follows:

1. Apply a vacuum of -0.1 MPa using a vacuum pump for 2 h.
2. Saturate the core with simulated water through self-absorption for 4 h, obtaining the pore volume V_P.
3. Perform oil displacement with water using an ISCO pump at a speed of 0.5 mL/min. Once oil appears at the outlet, increase the flow rates to 1 mL/min and 2 mL/min in sequence. The total injected oil volume is twice the pore volume, and the initial oil saturation S_{oi} is calculated.
4. Age the core for 3 days.
5. Connect the devices, fill the piston container with water and the polymer agent system, purge air from the pipelines and valves, and perform a pressure leak test at 2 MPa.
6. Then, proceed with the displacement experiment according to the experimental plan. Conduct water flooding. Upon reaching the predetermined water cut level, inject a slug of the designed chemical agent. Finally, conduct the subsequent water flooding until the composite water cut reaches 98%, at which point the experiment is terminated.
7. The process is conducted at the experimental temperature, monitoring pressure, and liquid production during displacement.
8. Data processing: calculate parameters such as fractional flow, oil recovery factor, and the equilibrium displacement degree using the following formulas.
   Fractional flow:

$$f_w = Q_i / \sum_1^n Q_i \times 100\% \tag{4}$$

Oil recovery factor:

$$R = \sum_1^n V_{oi} / \left(V_p \times S_{oi}\right) \times 100\% \tag{5}$$

Equilibrium displacement degree was calculated using Equation (3), where Q_i represents the instantaneous liquid production of each layer, mL; i denotes the ith layer; and n is the total number of layers; V_{oi} is the cumulative oil production of each layer, mL; V_p is the pore volume, mL; and S_{oi} is the oil saturation, %.

(1) Validation of Sweep Efficiency

The sweep efficiency is defined in Section 2.3.4. Here, two sets of three-tube parallel core displacement experiments are designed. The displacement flow chart are shown in Figure 2. By comparing the recovery factors and sweep efficiency indicators in two-stage discontinuous chemical flooding and single polymer slug flooding, the feasibility of the sweep efficiency definition and its sensitivity to the experimental scheme were determined. The specific experimental plan is shown in Table 3.

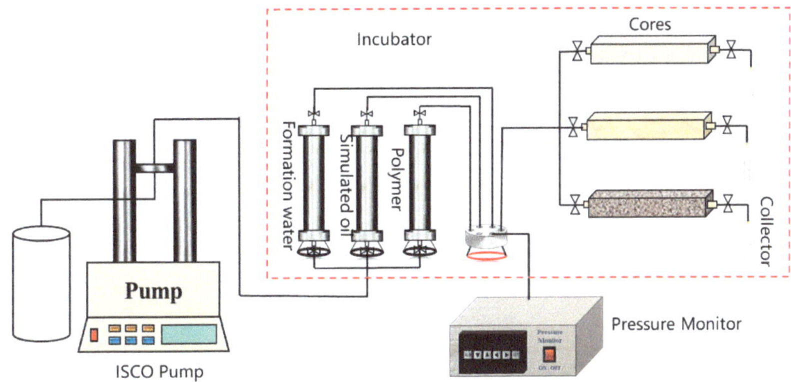

Figure 2. Experimental flow chart of the heterogeneous core model.

Table 3. Experimental scheme for the verification of displacement equilibrium degree.

No.	Core Permeability/mD	Transfer Timing	Experimental Scheme	Experimental Procedure
1	500/2000/5000	80%	The 0.7 PV polymer	Water flooding to a comprehensive water cut of 80%; polymer flooding of 0.7 PV after a water comprehensive water cut of 98%
2	500/2000/5000	80%	0.3 PV gel + 0.4 PV polymer	

(2) Design of Slug Size

The total size of the discontinuous chemical flooding slug was optimized using a three-plate parallel model. Electrodes and pressure points were arranged in the model. The physical model and displacement flow chart are shown in Figure 3. By injecting a large slug of the polymer system, the residual oil distribution characteristics were clarified using saturation electrodes, optimizing the best slug size. The specific experimental plan is shown in Table 4.

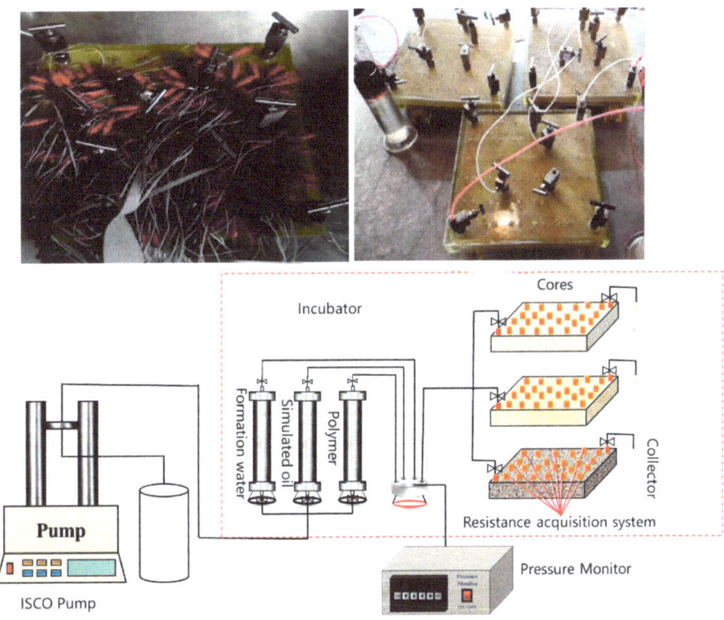

Figure 3. Physical diagram and flow diagram of the three-core large parallel model with electrodes.

Table 4. Experimental scheme of polymer slug size.

No.	Core Permeability/mD	Slug Size/PV	Experimental Scheme	Transfer Timing
1	500/2000/5000	1	Water flooding + 1 PV polymer + back water	Water cut of 80%

(3) Optimization of Combination Methods.

Based on the optimized conformance control system intensity, three discontinuous displacement combination methods were designed, including three-stage combinations, namely ① 0.2 PV gel + 0.2 PV polymer A + 0.3 PV polymer (strong–medium–weak), ② 0.3 PV polymer + 0.2 PV polymer A + 0.3 PV gel (weak–medium–strong), and ③ 0.2 PV polymer A + 0.2 PV gel + 0.3 PV polymer (medium–strong–weak), and a four-stage combination, namely ④ 0.2 PV polymer A + 0.1 PV gel + 0.15 PV polymer + 0.1 PV gel + 0.15 PV polymer (medium–strong–weak–strong–weak). These methods were evaluated through three-tube parallel displacement experiments (Figure 2) combined with sweep efficiency indicators, and the optimal system combination method was selected. The specific experimental plan is shown in Table 5.

Table 5. Optimization design of oil displacement system combination methods.

No.	Core Permeability/mD	Slug Size	Combination Mode	Experimental Scheme	Transfer Timing
1	500/2000/5000	0.7	Strong–medium–weak	0.2 PV gel + 0.2 PV polymer A + 0.3 PV polymer + post-water	Water cut of 80%
2	500/2000/5000	0.7	Weak–medium–strong	0.3 PV polymer + 0.2 PV polymer A + 0.3 PV gel + post-water	Water cut of 80%
3	500/2000/5000	0.7	Medium–strong–weak	0.2 PV polymer A + 0.2 PV gel + 0.3 PV polymer + post-water	Water cut of 80%
4	500/2000/5000	0.7	Medium–strong–weak–strong–weak	0.2 PV polymer A + 0.1 PV gel + 0.15 PV polymer + 0.1 PV gel + 0.15 PV polymer + post-water	Water cut of 80%

(4) Optimization of the Best Conversion Timing

Based on the optimal system combination method, the conversion timing of each discontinuous chemical flooding system was studied. Similarly, using the three-tube parallel displacement experiment (Figure 2) combined with sweep efficiency indicators, the effect of improving reservoir heterogeneity at four different conversion timings was evaluated, determining the best combination method. The specific experimental plan is shown in Table 6.

Table 6. Optimization design of the shift timing for oil displacement experiments.

No.	Core Permeability/mD	Slug Size	Experimental Scheme	Transfer Timing
1	500/2000/5000	0.7	0.2 PV polymer A + 0.1 PV gel + 0.15 PV	Water cut of 0%
2	500/2000/5000	0.7	polymer + 0.1 PV gel + 0.15 PV polymer	Water cut of 30%
3	500/2000/5000	0.7	+ back water	Water cut of 70%
4	500/2000/5000	0.7	(Medium–strong–weak–strong–weak)	Water cut of 80%

2.3.6. Verification of the Effectiveness of Discontinuous Chemical Flooding

To verify the beneficial effects of the discontinuous chemical flooding method optimized in Section 2.3.5, a five-tube parallel displacement experiment was conducted, comparing the discontinuous chemical flooding with continuous polymer flooding. The experimental steps followed those described in Section 2.3.5, and the experimental plan is outlined in Table 7.

Table 7. Five-core parallel displacement experiment schemes.

No.	Core Permeability/mD	Experimental Scheme	Experimental Procedure
1	500/2000/5000/7500/10,000	0.2 PV polymer A + 0.1 PV gel + 0.15 PV polymer + 0.1 PV gel + 0.15 PV polymer + post-water	Water flooding to a comprehensive water cut of 80%; polymer flooding post water to a total water cut of 100%
2	500/2000/5000/7500/10,000	The 0.7 PV polymer	

3. Results and Discussion

3.1. Performance Evaluation of the Oil Displacement Systems

3.1.1. Optimization of Gel System Formulation

The formulation design for gel application mainly included the design of polymer concentration and polymer-to-crosslinker ratio. Gel solutions were prepared using polymer concentrations of 1000, 1200, 1400, 1600, and 1750 mg/L, with a polymer-to-crosslinker ratio of 2:1. These solutions were then aged at 60 °C for a specified duration. The initial gelation time was defined as the intersection of the rapidly rising viscosity trend line with the x-axis. The stable gelation time was defined as the intersection of the trend line, where viscosity tended to stabilize with the rapidly rising trend line.

The results reveal that the solution with a concentration of 1750 mg/L underwent discoloration to yellow over time and exhibited a high gel strength. The results show that no good gel formation occurred at concentrations below 1600 mg/L, while, at 1750 mg/L, the gel strength was high, making it suitable for underground crosslinking to block high-permeability layers. Therefore, the optimal polymer concentration for the gel was between 1600–1750 mg/L.

A polymer with a concentration of 1600 mg/L was mixed with crosslinking agents at mass ratios of 4:1, 2:1, 1:1, 1:2, and 1:4 to obtain gel solutions. The viscosity curves were tested at 60 °C, and the experimental results are shown in Figure 4.

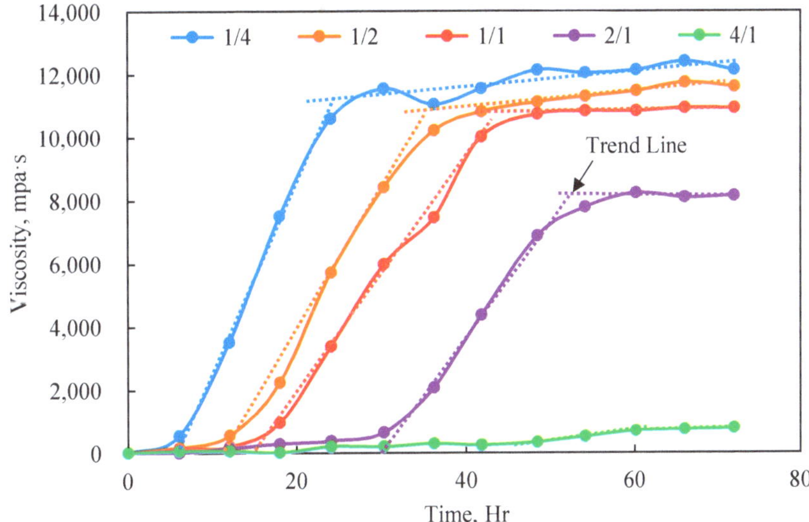

Figure 4. Viscosity curves of different polymer-to-crosslinker ratio systems.

Figure 4 shows that the ratio of polymer to crosslinking agent significantly affected gel formation. A decreased polymer-to-crosslinker ratio, indicating an increased amount of crosslinking agent, enhanced the collision probability between the polymer and the crosslinking agent, leading to a reduced gelation time. Additionally, the quantity of gel formed increased, causing gel properties to progressively dominate the system and highlighting the difference in viscosity before and after gelation. Conversely, a high polymer-to-crosslinker ratio signifies a lesser amount of crosslinking agent, insufficient for complete crosslinking of the polymers in solution. This leads to partial crosslinking and the formation of a relatively small amount of gel. Consequently, the system primarily exhibited polymeric properties, resulting in a lower viscosity and an extended crosslinking time. Hence, the viscosity change in the system before and after gelation was insignificant. A polymer-to-crosslinker ratio of 1:1 or 1:2 yielded a suitable gelation time and gel strength, making this range the preferred one. Gel formation commenced after 12 h of aging, with a gelation rate of approximately 300 mPa·s/h, and the final gel strength could reach 10,000 mPa·s.

The viscoelasticity of the gel with a concentration of 1600 mg/L and a polymer-to-crosslinker ratio of 1:2 was tested, and the experimental results are shown in Figure 5. The results indicate that, after gel formation, across various vibration frequencies, the elastic modulus (G′) consistently exceeded the viscous modulus (G″), indicating a predominantly elastic system. Specifically, it was found that the elastic modulus G′ was less than 1 Pa. According to industry standards, this gel was classified as a weak gel.

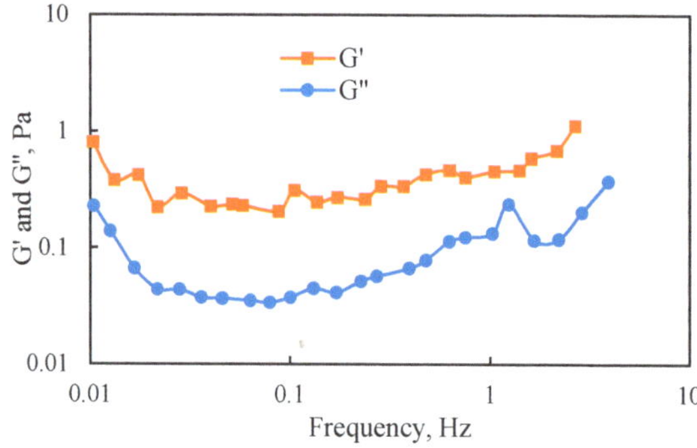

Figure 5. Relationship between viscoelastic moduli and vibration frequency of the gel under the optimal formula.

3.1.2. Graded Evaluation of the Performance of Oil Displacement Systems

The viscosity of polymer A and polymer B solutions at different concentrations was measured to determine their thickening properties, as shown in Table 8. The results indicate that the viscosity of the gel, as well as that of the polymer A and polymer B solutions, exhibited a stepwise variation. The viscosity of the gel was much higher than that of the polymer solutions, while the thickening ability of polymer A was stronger than that of regular polymers. As a result, the plugging strength of the three systems was classified into strong, medium, and weak levels.

Table 8. Viscosity range of polymer agents.

No.	Polymer Agents	Main Control Parameters	Scope	Intensity
1	Gel	Dynamic viscosity	>2000	Strong
2	Polymer A	Dynamic viscosity	200–350 cP	Medium
3	Polymer B	Dynamic viscosity	30–50 cP	Weak

3.2. Feasibility Verification of Displacement Equilibrium Evaluation Index

Figure 6 shows the displacement equilibrium curves for two-stage discontinuous displacement and continuous displacement. The overall curves exhibit a stepwise change: initially, only the high-permeability layers were activated, resulting in a control coefficient of 1/3, which caused the first point to be significantly lower. As displacement progressed, the medium- and low-permeability layers were subsequently activated, with a control coefficient of 1, leading to a jump in the second point. Observing Figure 6, the equilibrium displacement degree did not show a significant increase in the water flooding scheme, yielding the smallest final equilibrium displacement degree and, consequently, the smallest corresponding recovery degree. In contrast, in the polymer flooding scheme, the equilibrium displacement degree exhibited an increase at an injection volume of 0.4 PV. The increase was because the chemical flooding started to take effect. During the DCF process, the equilibrium displacement degree showed two distinct steps at injection volumes of 0.5 PV and 1.1 PV. The injection of the second slug in DCF resulted in the reuse of the reservoir.

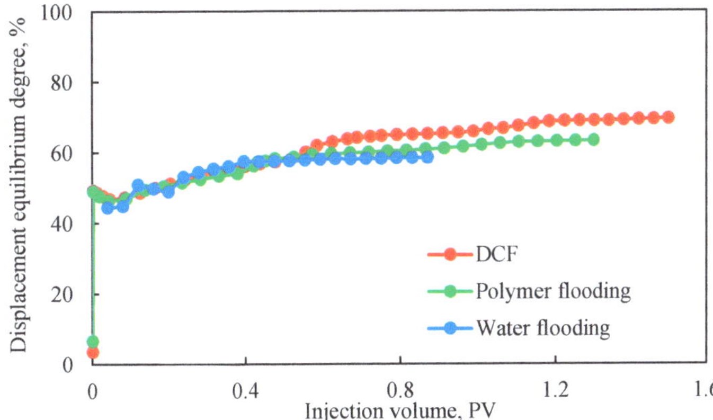

Figure 6. Comparison of discontinuous displacement and continuous flooding using displacement equilibrium degree.

Table 9 presents key parameters from the three displacement experiments, revealing that discontinuous displacement can increase recovery by more than 8% compared to continuous polymer flooding, demonstrating significant enhancement effects. When comparing the final recovery and displacement equilibrium among the three experiments, it becomes evident that the recovery factor showed a more noticeable numerical difference than the displacement equilibrium, making it easier to compare the pros and cons of different schemes.

Table 9. Comparison of displacement equilibrium degree and oil recovery.

	Combined Recovery at a Water Cut of 80%, %	Ultimate Recovery, %	Final Displacement Equilibrium, %
Water drive	22.24	39.17	58.49
DCF	23.18	58.21	69.46
Polymer flooding	23.59	46.74	63.23

However, the displacement equilibrium index defined showed the activation of each layer through stepwise changes. As the number of small layers increased and the reservoir heterogeneity worsened, this advantage became more pronounced, making it suitable for the combined development of multilayer thick oil reservoirs offshore. Although the recovery factor showed more significant differences, Table 8 also shows that the displacement equilibrium still had noticeable differences, though less obvious than the recovery factor, and it can still be used to compare different schemes. Therefore, the displacement equilibrium parameter proposed can be used for the subsequent optimization of discontinuous chemical flooding schemes.

3.3. Combination System Slug Size Design

Figure 7 shows the oil saturation field maps at different injection pore volumes during the oil displacement experiment, as described in Table 4. At 0.2 PV, the oil saturation at the outlet end of the high-permeability layer was higher compared to that at 0.1 PV. In the medium-permeability layer at 0.2 PV, the oil saturation in the main streamline exhibited a trend of increasing, slightly decreasing, and then increasing again. This trend was attributed to the accumulation of crude oil during the chemical flooding process and its gradual displacement. After 0.4 PV, dominant channels gradually formed in the high-permeability layer. Consequently, the average residual oil saturation in the high, medium, and low-permeability layers at 0.6 PV showed minimal variation compared to that at 0.8 PV, with a variation of less than 5‰. This was due to the significant development potential and high displacement efficiency observed in the early stages of polymer solution injection. However, with continued injection, the advantage of high-permeability channels became more pronounced, greatly reducing the potential sweep volume. Therefore, there was little difference in the remaining oil saturation field after 0.6 PV.

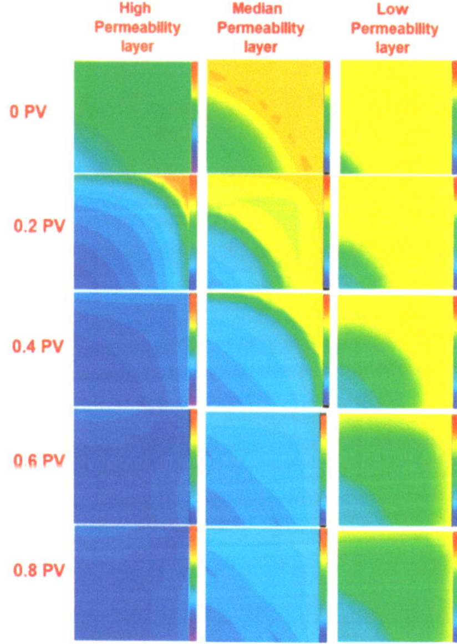

Figure 7. Oil saturation distribution diagram at different shift times of the combined system.

Considering that the discontinuous chemical flooding extended the effective displacement period, a slug size of 0.7 PV was chosen as the optimal injection volume for subsequent investigations. This selection maximizes the recovery improvement while mitigating resource waste.

3.4. Optimization of the Combination Method for the Profile Control System

Figure 8 shows the displacement equilibrium degree curves under four different discontinuous chemical flooding combination methods. During the water flooding phase, the displacement equilibrium degree remained stable at around 50%. After the injection of the polymer solution, the displacement equilibrium degree rapidly increased and then stabilized again. Notably, each transition to a new polymer slug resulted in a discernible elevation in the displacement equilibrium degree.

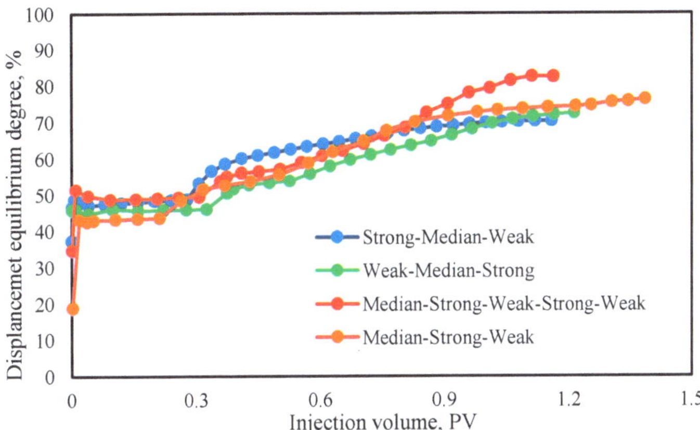

Figure 8. Displacement equilibrium degree curves for different discontinuous combination methods.

For reservoirs with identical physical properties, they tended to have similar waterflooding development stages, including the affected formations and recovery degrees. Therefore, the displacement equilibrium degree during the waterflooding phase tended to be consistent among these reservoirs.

However, different chemical systems exhibited varying injectability in different formations, resulting in differential impacts on those formations when injected. Consequently, the displacement equilibrium degree exhibited varying trends.

For the strong–medium–weak combination, as the slug intensity decreased, the increase in the displacement equilibrium degree gradually weakened, due to the high-saturation oil wall formed by the strong slug during the initial stages of chemical flooding which hindered subsequent medium and weak slugs to re-coalesce the oil wall. Additionally, as production progressed, the remaining oil saturation gradually decreased, further complicating coalescence.

The displacement equilibrium degree for the medium–weak–strong combination exhibited two distinct steps. Initially, the chemical flooding coalesced the residual oil left by waterflooding, but, due to the weaker intensity of the chemical system, the growth rate was slower than that of the strong–medium–weak combination. Subsequently, the injection of the strong slug further enhanced the swept volume and coalesced the remaining oil. As production continued and the slug intensity decreased, the displacement equilibrium degree gradually stabilized.

Similarly, the weak–medium–strong combination also showed two distinct steps in the displacement equilibrium degree. Due to the different timing of slug injection, the results and growth rates of the displacement equilibrium degree differed from those of the medium–weak–strong combination.

Observing the medium–strong–weak–strong–weak combination, the displacement equilibrium degree exhibited three distinct steps. This characteristic was due to the secondary plugging of the high-permeability layer with a strong slug prior to breakthrough which facilitated the further development of the medium- and low-permeability layers, further increasing the displacement equilibrium degree.

By comparing the results of Figures 6 and 8, the final displacement equilibrium degrees followed this order: five-stage > three-stage > two-stage discontinuous chemical flooding. Thus, the "medium-strong-weak-strong-weak" combination achieved the highest displacement equilibrium degree, indicating better displacement performance. The primary reason was that the five-stage displacement method advanced over the three-stage method by using smaller slug sizes of gel to effectively block high-permeability layers. This was subsequently followed by the utilization of polymer systems to efficiently displace oil from medium- and low-permeability layers. This approach avoids the problems of high pressure, low-permeability layer contamination, and long-term polymer channeling that could occur with large gel slugs.

The "strong-medium-weak" method resulted in the lowest displacement equilibrium degree, mainly because the gel's blocking strength was too high, and injecting gel too early would hinder the full utilization of the remaining oil in the high-permeability layers.

The "medium-strong-weak-strong-weak" combination leveraged polymer A to fully displace high-permeability layers before using the gel to block them, allowing for the effective utilization of medium- and low-permeability layers. The alternation between strong and weak displacement helps avoid abnormal pressure increases caused by large gel slugs. Therefore, the optimal discontinuous chemical flooding combination was the "medium-strong-weak-strong-weak" configuration.

3.5. Optimization of the Best Switching Timing for the Profile Control System Combination

Figure 9 displays the displacement equilibrium degree curves under five distinct switching timings. The primary distinction among the various design schemes lied in the timing of chemical injection which resulted in varying residual oil saturation levels after waterflooding. However, the chemical systems and slug sizes remained consistent, leading to similar shapes in the displacement equilibrium degree curves. However, the rates of increase in the displacement equilibrium degree varied among different injection timings, due to the different distributions and saturation states of residual oil under different injection timings which affected the interaction and coalescence between the chemical agents and the crude oil. It can be observed that the highest displacement equilibrium degree, and, thus, the best displacement performance, occurred when the switching timing was at a 70% water cut. Conversely, the lowest displacement equilibrium degree, and the worst oil recovery factor, was observed when the switching occurred at a 0% water cut.

Switching to polymer flooding too early means that the full potential of water flooding is not exploited, resulting in a high oil saturation in the reservoir. This causes a rapid increase in injection pressure. Once channeling occurs in the high-permeability layers, mobilizing medium- and low-permeability layers becomes even more challenging. Additionally, switching too early significantly increases production costs. Conversely, switching too late results in a low oil saturation in the reservoir, with residual oil forming as oil films in small pores, making it difficult for the injected polymer solution to displace the oil.

Therefore, switching the polymer flooding system too early or too late is not beneficial for improving oil recovery. The optimal switching timing for discontinuous polymer flooding should be when the water cut is between 70% and 90%.

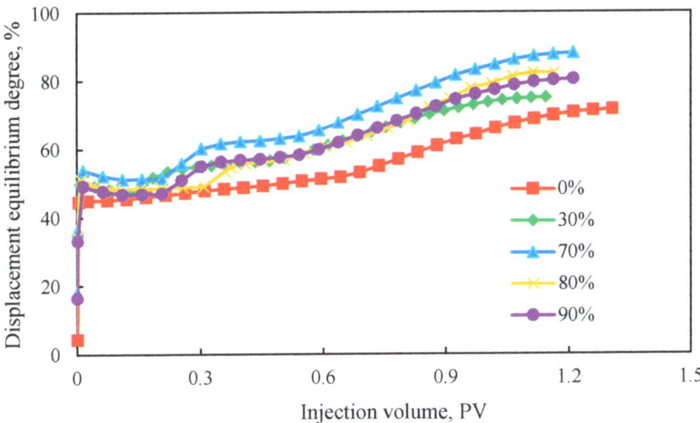

Figure 9. Displacement equilibrium degree curves at different injection times.

3.6. EOR Effect of Discontinuous Chemical Flooding

Figure 10 compares the recovery factor of various layers in the five-tube parallel flooding experiments, utilizing discontinuous chemical flooding (DCF) and continuous chemical flooding (CF). The DCF achieved a final oil recovery factor of 51.47%, representing an increase of 24.55% compared to CF. This enhancement was primarily attributed to improved mobilization in the medium- and low-permeability layers. CF primarily mobilized high and secondary high-permeability layers, whereas DCF not only enhanced the recovery in high-permeability layers, but also significantly improved the mobilization in medium- and low-permeability layers. Notably, in DCF, the recovery factor of the medium-permeability layer was lower than that of the second-lowest permeability layer, due to gel contamination. This indicates that stronger gel plugging should only be applied after sufficient mobilization of the medium-permeability layers to prevent channeling.

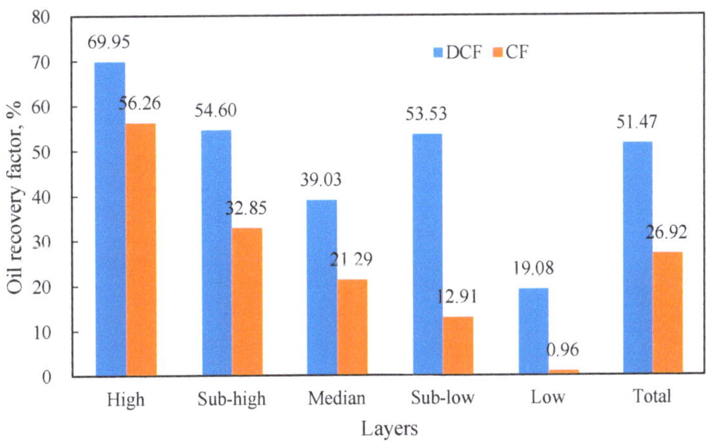

Figure 10. Comparison of recovery factors for different layers between DCF and CF in the five-core parallel flooding experiment.

Analyzing the above in conjunction with Figure 11, during the DCF process, the primary production phase for the sub-high permeability layer occurred during the initial slug injection. However, this initial strong slug effectively sealed off the sub-high permeability layer, resulting in suboptimal subsequent production. Conversely, the injection of two strong slugs blocked higher permeability layers, providing the sub-low permeability layer with two production opportunities. Consequently, the oil recovery of both layers is comparable.

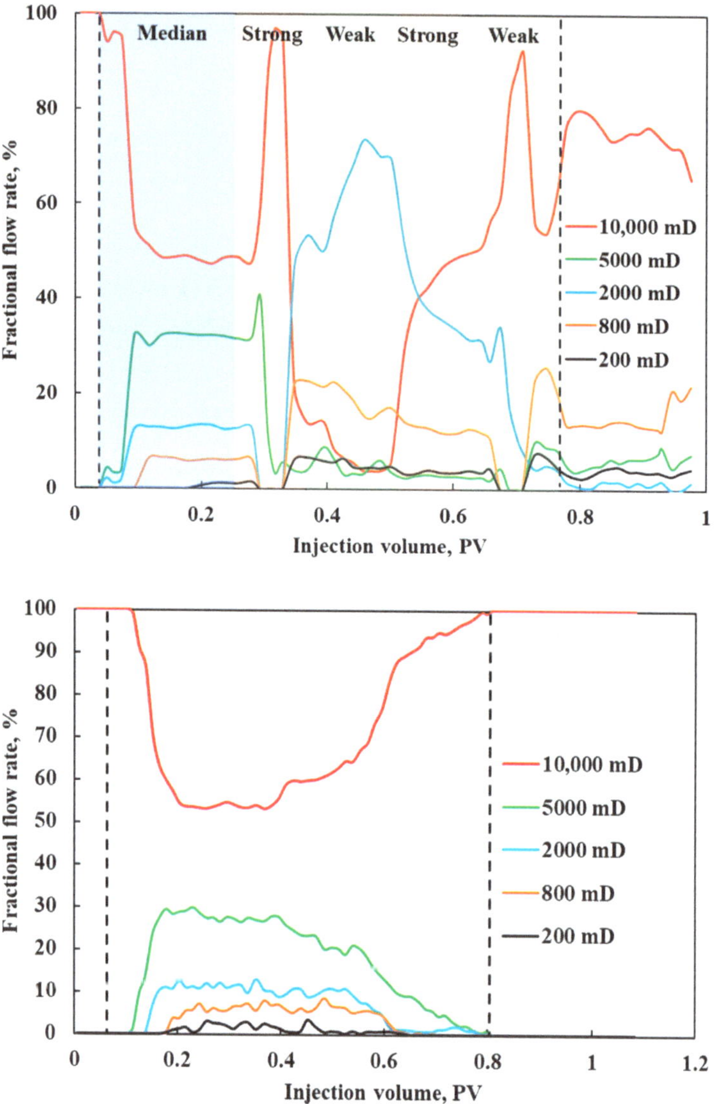

Figure 11. Comparison of fractional flow rate distribution between DCF and CF in five-core parallel flooding experiment.

Figure 11 compares the flow rate allocation between DCF and CF for each layer in the five-tube parallel flooding experiment. During the initial injection of polymer A, the flow rate in the high-permeability layer decreased to around 50%, enhancing mobilization in the medium- and low-permeability layers. However, continuing injection of a single slug system could revert the profile, with the high-permeability layer's flow rate significantly increasing again. After the first gel slug injection in DCF, the flow rate in the medium-permeability layer improved significantly, but it subsequently fell below that of the second-lowest permeability layer. This observation supports the hypothesis of gel contamination in the medium-permeability layer. DCF effectively optimized the flow rate curves at each stage. Although no oil was ultimately recovered from any layer, the high-permeability layer's flow rate remained below 80%, demonstrating its sustained effectiveness. The improvement of the injection and production profile in heterogeneous reservoirs is the key reason for the enhanced oil recovery achieved by DCF.

Upon injection of the chemical system in CF, the flow rate of the high-permeability layer decreased to approximately 50%. However, this effect was not sustained for a long period. The flow rate of the high-permeability layer significantly increased again at 0.4 PV, while the flow rates of the other layers remained below 15%. Based on this analysis, it is concluded that, in strongly heterogeneous multilayer reservoirs, the effectiveness of CF in improving the water injection profile is limited.

Figure 12 presents a comparison of the displacement equilibrium degree for DCF and CF. The stair step pattern in the displacement equilibrium was more pronounced in the five-tube parallel flooding experiment. These curves distinctly showcase the activation of each layer. Initially, both DCF and CF exhibited similar displacement equilibrium degrees. However, during the continuous injection of polymer A, CF's equilibrium degree increased slowly, whereas DCF's equilibrium degree improved significantly. The observed decline in DCF's displacement equilibrium stemmed from the widening gap between the overall recovery factor and the maximum recovery factor which is linked to the rapid mobilization of the high-permeability layer. The displacement equilibrium degree was influenced by two factors: the number of mobilized layers and the variations in mobilization among these layers. Throughout the CF process, the displacement equilibrium was clearly delineated into five stages, aligning with the data presented in Figure 11. These stages represent the sequential utilization of the five reservoir layers, leading to a gradual increase in the displacement equilibrium. As the displacement process progressed, the equilibrium exhibited a gradual increase before leveling off. This was mainly due to the oil recovery of the high-permeability layer, namely the maximum oil recovery, reaching over 50%, thereby constraining a further increase. Simultaneously, the growth rate of the comprehensive oil recovery also gradually decreased. These two factors interact, collectively shaping the final contour of the displacement equilibrium curve.

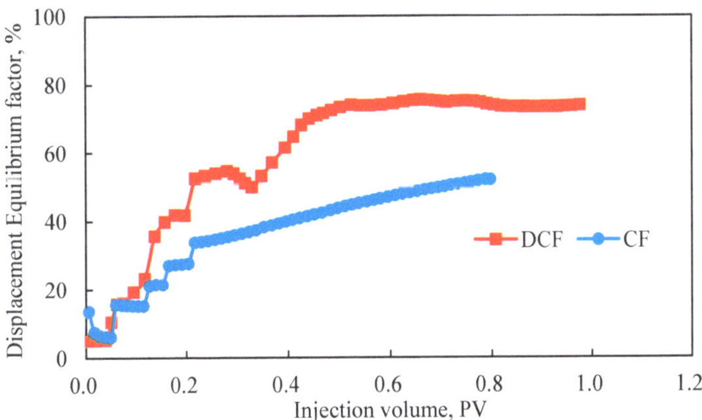

Figure 12. Comparison of displacement equilibrium degree between DCF and CF in five-core parallel flooding experiments.

Ideally, all layers in DCF are mobilized (with the control coefficient of 1) and exhibit uniform recovery factors (with the homogeneity coefficient of 1), achieving a maximum equilibrium degree of 100%. Figure 12 affirms that displacement equilibrium serves as a valuable tool for assessing the efficacy of various displacement schemes and underscores the notable advantages of DCF.

4. Conclusions

The profile control and flooding performance of the gel, polymer A, and polymer B are evaluated based on parameters like viscosity and viscoelasticity, categorizing them into strong, medium, and weak grades. This study emphasizes the evaluation criteria for discontinuous chemical flooding (DCF) and the optimization of injection parameters through laboratory experiments. A displacement equilibrium degree is defined, considering the number and mobilization efficiency of mobilized layers. The slug size, combination strategy, and shift timing of DCF are optimized using the parallel core flooding experiments. The benefits of DCF in oil displacement are validated through five-core parallel flooding experiments. The specific conclusions are as follows:

1. The gel formula is optimized with a polymer concentration of 1600 mg/L and a polymer-to-gel ratio of 1:2. The gel exhibits strong viscoelasticity, with its elastic modulus significantly exceeding its viscous modulus. With a viscosity of 10,000 mPa·s, the gel is classified as weak. The profile control and flooding intensities of the three systems are rated as follows: gel—strong, polymer A—medium, and polymer B polymer—weak.

2. The displacement equilibrium degree reflects the mobilization efficiency and timing of each layer during DCF. The displacement equilibrium degree curve presents a stepwise pattern. The efficiency of DCF becomes more significant with the number of thin layers increasing and heterogeneity intensifying. This metric is highly sensitive to different flooding strategies, effectively highlighting differences among each displacement scheme.

3. Under the combination of a permeability of 5000/2000/500 mD, the optimal dosage for chemical discontinuous flooding is 0.7 PV. The optimal combination pattern is the injection in the sequence of "medium-strong-weak-strong-weak". This pattern can achieve a displacement equilibrium degree of 82.3%. The optimal shift timing of polymers injection occurs at a water cut of 70%, yielding a displacement equilibrium degree of 87.7%.

4. In comparison to CF, DCF demonstrates a higher incremental oil recovery of 24.5% in the five-core parallel displacement experiment. DCF effectively blocks dominant flow channels and controls injection pressure, with its advantages being more significant in heterogeneous reservoirs.

Author Contributions: Conceptualization, X.L. and J.Z.; Methodology, X.L. and Z.L.; Formal analysis, K.H.; Investigation, Y.Z. and C.G.; Resources, R.X.; Data curation, J.Z., C.G. and K.H.; Writing—original draft, X.L.; Writing—review & editing, Y.Z., Z.L. and R.X.; Supervision, J.Z., Z.L. and Y.L.; Project administration, J.Z. and Y.L.; Funding acquisition, J.Z. and Y.L. All authors have read and agreed to the published version of the manuscript.

Funding: This research was funded by the Major Science and Technology Project of CNOOC (No. KJGG2021-0504) and the National Natural Science Foundation of China (Grant 52074318) for the support of this work.

Data Availability Statement: The original contributions presented in this study are included in the article. Further inquiries can be directed to the corresponding author.

Conflicts of Interest: Authors Xianjie Li, Jian Zhang, Cuo Guan, Ke Hu, Ruokun Xian were employed by the company CNOOC Research Institute Co., Ltd. The remaining authors declare that the research was conducted in the absence of any commercial or financial relationships that could be construed as a potential conflict of interest.

References

1. Steinmann, S.; Gronski, W.; Friedrich, C. Quantitative rheological evaluation of phase inversion in two-phase polymer blends with cocontinuous morphology. *Rheol. Acta* **2002**, *41*, 77–86. [CrossRef]
2. Pereborova, N.V. Qualitative Assessment of Deformation-Performance Properties of Polymer Textile Materials. *Fibre Chem.* **2023**, *55*, 146–148. [CrossRef]
3. Afolabi, R.O.; Oluyemi, G.F.; Officer, S.; Ugwu, J.O. Hydrophobically associating polymers for enhanced oil recovery—Part A: A review on the effects of some key reservoir conditions. *J. Pet. Sci. Eng.* **2019**, *180*, 681–698. [CrossRef]
4. Mahajan, S.; Yadav, H.; Rellegadla, S.; Agrawal, A. Polymers for enhanced oil recovery: Fundamentals and selection criteria revisited. *Appl. Microbiol. Biotechnol.* **2021**, *105*, 8073–8090. [CrossRef]
5. Chang, H.L. Polymer flooding technology yesterday, today, and tomorrow. *J. Pet. Technol.* **1978**, *30*, 1113–1128. [CrossRef]
6. Sheng, J.J.; Leonhardt, B.; Azri, N. Status of Polymer-Flooding Technology. *J. Can. Pet. Technol.* **2015**, *54*, 116–126. [CrossRef]
7. Tahir, M.; Hincapie, R.E.; Langanke, N.; Ganzer, L.; Jaeger, P. Coupling Microfluidics Data with Core Flooding Experiments to Understand Sulfonated/Polymer Water Injection. *Polymers* **2020**, *12*, 1227. [CrossRef]
8. Guo, H.; Song, K.; Liu, S.; Zhao, F.; Wang, Z.; Xu, Y.; Liu, J.; Tang, E.; Yang, Z. Recent Advances in Polymer Flooding in China: Lessons Learned and Continuing Development. *SPE J.* **2021**, *26*, 2038–2052. [CrossRef]
9. Li, X.; Zhang, F.; Liu, G. Review on polymer flooding technology. *IOP Conf. Ser. Earth Environ. Sci.* **2021**, *675*, 012199. [CrossRef]
10. Mohsenatabar Firozjaii, A.; Saghafi, H.R. Review on chemical enhanced oil recovery using polymer flooding: Fundamentals, experimental and numerical simulation. *Petroleum* **2020**, *6*, 115–122. [CrossRef]
11. Kakati, A.; Kumar, G.; Sangwai, J.S. Low Salinity Polymer Flooding: Effect on Polymer Rheology, Injectivity, Retention, and Oil Recovery Efficiency. *Energy Fuels* **2020**, *34*, 5715–5732. [CrossRef]
12. Obino, V.; Yadav, U. Application of Polymer Based Nanocomposites for Water Shutoff—A Review. *Fuels* **2021**, *2*, 304–322. [CrossRef]
13. Chen, X.; Li, Y.; Liu, Z.; Zhang, J.; Chen, C.; Ma, M. Investigation on matching relationship and plugging mechanism of self-adaptive micro-gel (SMG) as a profile control and oil displacement agent. *Powder Technol.* **2020**, *364*, 774–784. [CrossRef]
14. Liu, C.; Zhou, W.; Jiang, J.; Shang, F.; He, H.; Wang, S. Remaining Oil Distribution and Development Strategy for Offshore Unconsolidated Sandstone Reservoir at Ultrahigh Water-Cut Stage. *Geofluids* **2022**, *2022*, 6856298. [CrossRef]
15. Vasiltsova, V.M. Problems development of the offshore oil and gas fields deposits. *J. Min. Inst.* **2016**, *218*, 345.
16. Zheng, A.; Feng, Q.; Wei, Q.; Liu, D. Research on profile inversion pattern of polymer flooding. *Geosyst. Eng.* **2018**, *21*, 135–141. [CrossRef]
17. Tahir, M.; Hincapie, R.E.; Ganzer, L. An Elongational and Shear Evaluation of Polymer Viscoelasticity during Flow in Porous Media. *Appl. Sci.* **2020**, *10*, 4152. [CrossRef]
18. Wang, D.; Dong, H.; Lv, C.; Fu, X.; Nie, J. Review of practical experience by polymer flooding at Daqing. *SPE Reserv. Eval. Eng.* **2009**, *12*, 470–476. [CrossRef]
19. Liao, G.; Wang, Q.; Wang, H.; Liu, W.; Wang, Z. Chemical flooding development status and prospect. *Acta Pet. Sin.* **2017**, *38*, 196.
20. Liu, Z.; Li, Y.; Lv, J.; Li, B.; Chen, Y. Optimization of polymer flooding design in conglomerate reservoirs. *J. Pet. Sci. Eng.* **2017**, *152*, 267–274. [CrossRef]
21. Chen, X.; Li, Y.-Q.; Liu, Z.-Y.; Trivedi, J.; Gao, W.-B.; Sui, M.-Y. Experimental investigation on the enhanced oil recovery efficiency of polymeric surfactant: Matching relationship with core and emulsification ability. *Pet. Sci.* **2023**, *20*, 619–635. [CrossRef]
22. Shi, H.; Liu, Y.; Jiang, L.; Zheng, J.; Gan, L. Integrated Solutions for the Redevelopment Plan of Offshore Mature-Oilfield. In Proceedings of the International Petroleum Technology Conference, Virtual, 23 March–1 April 2021; IPTC: London, UK, 2021.
23. De Pasquale, G.; Bertana, V.; Scaltrito, L. Experimental evaluation of mechanical properties repeatability of SLA polymers for labs-on-chip and bio-MEMS. *Microsyst. Technol.* **2018**, *24*, 3487–3497. [CrossRef]
24. Bamzad, S.; Nourani, M.; Ramazani, S.A.A.; Masihi, M. Experimental Investigation of Flooding Hydrolyzed-Sulfonated Polymers for EOR Process in a Carbonate Reservoir. *Pet. Sci. Technol.* **2014**, *32*, 1114–1122. [CrossRef]
25. Góes, M.R.R.; Guedes, T.A.; Teixeira, R.G.; Melo, P.A.; Tavares, F.W.; Secchi, A.R. Multiphase flow simulation in offshore pipelines: An accurate and fast algorithm applied to real-field data. *Chem. Eng. Sci.* **2023**, *268*, 118438. [CrossRef]
26. Chen, X.; Zhang, Q.; Li, Y.; Liu, J.; Liu, Z.; Liu, S. Investigation on enhanced oil recovery and CO_2 storage efficiency of temperature-resistant CO_2 foam flooding. *Fuel* **2024**, *364*, 130870. [CrossRef]
27. Chen, X.; Li, Y.; Liu, Z.; Li, X.; Zhang, J.; Zhang, H. Core-and pore-scale investigation on the migration and plugging of polymer microspheres in a heterogeneous porous media. *J. Pet. Sci. Eng.* **2020**, *195*, 107636. [CrossRef]
28. Chen, X.; Li, Y.; Liu, Z.; Zhang, J.; Trivedi, J.; Li, X. Experimental and theoretical investigation of the migration and plugging of the particle in porous media based on elastic properties. *Fuel* **2023**, *332*, 126224. [CrossRef]
29. Zheng, A.; Liu, D.; Shao, Y. Research on profile inversion controlling methods of multi-slug polymer alternative injection. *Geosyst. Eng.* **2020**, *23*, 92–100. [CrossRef]

30. Zhu, Y.; Gao, W.; Li, R.; Li, Y.; Yuan, J.; Kong, D.; Liu, J.; Yue, Z. Action laws and application effect of enhanced oil recovery by adjustable-mobility polymer flooding. *Acta Pet. Sin.* **2018**, *39*, 189.

31. Zhang, J. Study on discontinuous chemical flooding model of offshore high water cut oilfield. *China Offshore Oil Gas* **2023**, *35*, 70–77.

32. Zhang, J.; Li, Y.; Li, X.; Guan, C.; Chen, X.; Liang, D. Mechanism of enhanced oil recovery by discontinuous chemical flooding in Bohai oilfield. *Acta Pet. Sin.* **2024**, *45*, 988.

33. Asadizadeh, S.; Ayatollahi, S.; Zarenezhad, B. Performance evaluation of a new nanocomposite polymer gel for water shutoff in petroleum reservoirs. *J. Dispers. Sci. Technol.* **2019**, *40*, 1479–1487. [CrossRef]

 *polymers*

Article

Study on Surfactant–Polymer Flooding after Polymer Flooding in High-Permeability Heterogeneous Offshore Oilfields: A Case Study of Bohai S Oilfield

Yingxian Liu [1], Lizhen Ge [1], Kuiqian Ma [1], Xiaoming Chen [1], Zhiqiang Zhu [1] and Jirui Hou [2,*]

[1] CNOOC China Limited Tianjin Branch, Tianjin 300459, China; gelzh2@cnooc.com.cn (L.G.); 18810390092@163.com (K.M.)
[2] Research Institute of Unconventional Petroleum Science and Technology, China University of Petroleum (Beijing), Beijing 102249, China
* Correspondence: houjirui@cup.edu.cn

Abstract: Polymer flooding is an effective development technology to enhance oil recovery, and it has been widely used all over the world. However, after long-term polymer flooding, a large number of oilfields have experienced a sharp decline in reservoir development efficiency. High water cut wells, serious dispersion of residual oil distribution and complex reservoir conditions all bring great challenges to enhance oil recovery. In this study, the method of enhancing oil recovery after polymer flooding was studied by taking the S Oilfield as an example. A surfactant–polymer system suitable for high-permeability heterogeneous oilfields was developed, comprising biogenic surfactants and polymers. Microscopic displacement experiments were conducted using cast thin sections from the S Oilfield, and nuclear magnetic resonance was employed for core displacement experiments. Numerical simulation experiments were also conducted on the S Oilfield. The results show that the enhanced oil recovery mechanism of the surfactant–polymer system is to adjust the flow direction, expand the swept volume, emulsify crude oil and reduce interfacial tension. Surfactant–polymer flooding proves to be effective in improving recovery efficiency, significantly reducing the time of flooding and further enhancing the strong swept area. The nuclear magnetic resonance results indicate a high amplitude of passive utilization of residual oil during the surfactant–polymer flooding stage, highlighting the enormous potential for an increased recovery ratio. Surfactant–polymer flooding emerges as a more suitable technique to enhance oil recovery in the post polymer-flooding stage in high-permeability heterogeneous oilfields.

Keywords: heavy oil reservoir; offshore; polymer flooding; surfactant–polymer flooding; enhanced oil recovery

Citation: Liu, Y.; Ge, L.; Ma, K.; Chen, X.; Zhu, Z.; Hou, J. Study on Surfactant–Polymer Flooding after Polymer Flooding in High-Permeability Heterogeneous Offshore Oilfields: A Case Study of Bohai S Oilfield. *Polymers* 2024, 16, 2004. https://doi.org/10.3390/polym16142004

Academic Editor: Marta Otero

Received: 8 May 2024
Revised: 6 June 2024
Accepted: 24 June 2024
Published: 12 July 2024

1. Introduction

In recent years, with the declining discovery of new petroleum reserves, enhancing oil recovery (EOR) technologies have become pivotal for meeting future energy demands [1]. Well-known chemical EOR techniques include polymer flooding, surfactant flooding, and alkaline flooding [2,3]. Among these, polymer flooding has been employed in the petroleum industry for over 40 years [4], demonstrating the potential to increase the recovery efficiency of geological reserves by 5–30% [5]. China, being the largest petroleum producer in chemical EOR projects, primarily relies on polymer flooding [6,7]. The first and second largest oilfields implementing polymer flooding in China are Daqing Oilfield and Shengli Oilfield. The oil production of these two oilfields mainly depends on polymer flooding [8–11]. Moreover, there is a growing emphasis on polymer flooding in heavy oil reservoirs and offshore oilfields.

While polymer flooding has been industrialized worldwide, its potential to enhance oil recovery in practical field applications is considerably limited [12]. The mechanism of polymer flooding involves increasing the viscosity of the displacing fluid, reducing the flowability contrast between the displacing fluid and crude oil, thereby mitigating the fingering effect [13]. The flowability of the displacing phase equals or is lower than that of the oil phase [14,15]. During the process of polymer flooding displacing crude oil, interactions such as electrostatic and London dispersion forces occur between polymer molecules and the rock surface [16]. Under conditions of high reservoir salinity and elevated formation temperatures, polymer viscosity loss occurs. These factors result in the final viscosity of the injected agent in the reservoir being lower than the targeted viscosity, thereby diminishing the effectiveness and efficiency of polymer flooding [17]. Due to the severe heterogeneity of the reservoir, immature well patterns, and high temperatures with elevated salinity, approximately half of the petroleum reserves remain in the reservoir after polymer flooding [18]. The current challenges with polymer flooding include limited improvement in sweep efficiency and low oil displacement efficiency [19]. Enhancing the sweep efficiency coefficient (also known as consistency control) is a prerequisite for increasing oil recovery after polymer flooding [20]. The continuous expansion of polymer injection in oilfields, along with the gradual increase in polymer solution concentration, has led to issues such as excessive injection pressure and injection difficulties in injection wells [21].

After polymer flooding refers to the late stage of polymer flooding with a high water cut, and Residual polymer makes it difficult to effectively displace crude oil during this stage. In order to enhance oil recovery in reservoirs after polymer flooding, a series of new technologies are put forward to improve oil recovery, such as profile control, water plugging, surfactant–polymer flooding (SP flooding), ternary compound flooding (ASP flooding) and foam composite flooding. A large number of studies have shown that profile control can effectively adjust the water absorption profile, and water plugging can effectively block the high-permeability layer, thus expanding the swept volume and improving the displacement effect. SP flooding, ASP flooding and foam composite flooding have different advantages and can improve the swept efficiency [22]. The investigated literature has studied the exploitation technology and difficulties of offshore oilfields. Shuang Liang, in order to improve the development effect of offshore oil fields at the high water cutting stage, studied the technology of nitrogen foam flooding and polymer microsphere flooding [23]. Lizhen Ge clarified the mechanism of weak gel flooding and remaining oil distribution in the LD Oilfield [24]. There are relatively few studies on the application of SP flooding technology in offshore oilfields. In order to study EOR methods after polymer flooding in offshore oilfields and explore new technologies to enhance oil recovery, the S Oilfield was taken as an example to carry out corresponding theoretical and experimental research in this paper.

The S Oilfield is situated offshore of the Bohai Bay Liaohe Basin, characterized by a cover-type semi-horst structure developed on the pre-Tertiary limestone basement, trending in a north-east direction with a simple structural configuration [25]. The sedimentary type is front-edge deposition of a river delta, predominantly featuring subaqueous distributary channels and estuarine dam sand bodies. The S Oilfield belongs to heavy oil reservoirs with significant differences in underground crude oil properties. The reservoir temperature of the S Oilfield is 65 °C, the reservoir pressure is 13.5 MPa, and the viscosity of crude oil is 24~452 mPa · s. The water content of the S Oilfield is 69.2%, and the recovery degree is 26.8%. Currently, the oilfield is in the late stage of polymer injection, characterized by a high water cut, and the remaining oil is primarily located in areas where polymer flooding has not penetrated or has low sweep efficiency, such as near faults [26]. The distribution of remaining oil is more scattered and complex, posing significant challenges to recovering these reserves in polymer-flooded blocks.

The S Oilfield faces several difficulties: Firstly, due to the complex offshore reservoir conditions with multiple vertical control layers in a single well, and limitations imposed by offshore platforms, implementing measures such as layer adjustments and well pattern improvements to address the oil–water issue is challenging. Secondly, the offshore platform lacks freshwater resources required for polymer injection, relying instead on formation water and seawater with elevated salinity. Thirdly, the S Oilfield has large well spacing, severe reservoir heterogeneity, and the polymer tends to channel into high-permeability layers, complicating the recovery of remaining oil in wells [23,24,27]. If a standalone polymer flooding method continues to be employed, it will adversely impact the development effectiveness of polymer flooding [28].

Currently, the enhanced oil recovery (EOR) technology after polymer flooding in offshore oilfields requires the integration of adjustment, plugging, and displacement techniques to expand the sweep volume, thereby enhancing the oil displacement efficiency. ASP compound flooding encounters challenges such as alkali scaling and corrosion [29]. Due to environmental and platform constraints in offshore oilfields, the difficulty in handling produced fluids is greater than in onshore fields. Therefore, under current technological conditions, ASP compound flooding is not suitable for use in offshore oilfields [30]. Due to limited space on offshore platforms, it is preferable to use the equipment and injection processes already employed in polymer flooding after polymer flooding. Considering the onsite conditions and production circumstances in offshore oilfields, the S Oilfield recommends the use of an SP system combining polymers and surfactants for EOR after polymer flooding. The surfactant–polymer system used in this study is composed of a biological surfactant and hydrophobically associated polymer. This kind of surfactant–polymer system, with a low concentration of surfactants and good ability to reduce interfacial tension, can be effectively applied to high-permeability heavy oil reservoirs in the Bohai Sea.

2. Methods and Materials

2.1. Surfactant–Polymer System

In the surfactant–polymer system, the surfactant system primarily comprises lipopeptide biosurfactants as the main agent, with sodium petroleum sulfonate as an auxiliary agent for compounding. In the biosurfactant system, the ratio of lipopeptide to the auxiliary agent is 1:1, with a total mass concentration of 0.34%. The polymer selected is a hydrophobically associated polymer. The polymer concentration is set at 3000 mg/L. Operating at 60 °C with an underground viscosity of 50 cP, the interfacial tension between the surfactant–polymer system and S crude oil can reach 1.1×10^{-3} mN/m.

The biosurfactant used is Surfactin, produced by Maclean, with a purity of 99%. It has a molecular weight of 1036.34, a boiling point of 1268.3 ± 65.0 °C and a density of 1.037 ± 0.06 g/cm^3. It is derived from sodium subtilis lipopeptide. The sodium petroleum sulfonate is also produced by Maclean, with a purity of 50%. The hydrophobically associated polymer is AP-P4. AP-P4 has good viscosity enhancement, shear resistance, temperature resistance up to 70 °C, salt resistance up to 10,000 mg/L, calcium and magnesium ion resistance up to 500 mg/L and good mobility control. The formation water has a mineralization level of 8000 mg/L, with calcium and magnesium ion concentrations of 240 mg/L.

The configuration method for the surfactant–polymer system is as follows:

(1) Weigh 500 mL of formation water and set the stirrer to a speed of 300 rpm for stirring. Weigh 6000 mg of the hydrophobically associated polymer and slowly add it to deionized water to avoid the occurrence of fisheye phenomena due to rapid addition. If fisheye occurs, reconfigure. After the polymer is dissolved, reduce the stirring speed to 80 to avoid damaging the polymer chain structure due to excessive speed.

(2) Weigh 500 mL of formation water, and set the stirrer to a speed of 180 rpm for stirring. Weigh 0.34 g of the main agent and 0.64 g of the auxiliary agent, add them separately to the formation water, and stir. After uniform stirring, add the biosurfactant system to the stirred polymer. Stir at low speed until the biosurfactant system and polymer are completely dissolved. The surfactant–polymer system configuration is complete.

2.2. Design and Fabrication of Micro-Physical Model

(1)   Similarity Criteria

Optimizing the design of indoor experimental conditions through similarity criteria aims to make the flow patterns in the model similar to the actual reservoir flow pattern. The primary similarity indicator is chosen to be motion similarity, matching the similarit of reservoir conditions with those of the S Oilfield. The specific results are shown in Table S1. The formula is as follows:

$$Re = \frac{\rho V L}{\mu}$$

where

ρ is the fluid density, measured in kg/m^3; V is the fluid velocity, measured in m/s; L is the characteristic length, measured in meters; and μ is the fluid viscosity, measured in Pa·s.

(2)   Model Design

The design and fabrication of the model involve using the actual core structure from the S Oilfield to closely simulate the geometric structure of the core. Initially, the real core is sliced, and these core slices are used to create cast thin sections. The actual structure of the cast thin sections is then outlined to represent the particle structure, and this is transformed into a CAD model design. Based on the model design, an accurate chemical etching method is employed on glass to create the core's pore structure, and the model is subsequently sealed. The resulting micro-scale visualization model is proportionally scaled to match the actual core dimensions, as depicted in Figure 1.

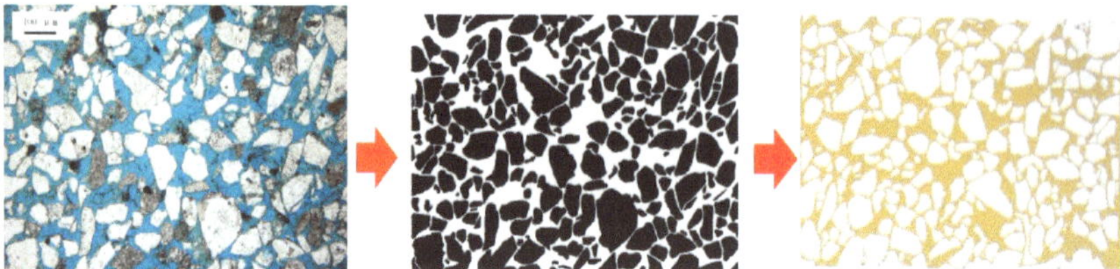

Figure 1. The process of making micro models for the S Oilfield.

(3)   Experimental Instruments and Procedures

The instruments used for microscopic displacement experiments include a dual-cylinder pump, intermediate container, six-way valve, pressure gauge, microscopic model holder, microscopic model imaging system, and connecting pipelines, as shown in Figure 2. The working pressure of the air pump ranges from 0 to 20 MPa, with a working flow rate of 0 to 15 mL/min. The volume of the intermediate container is 50 mL, with a working pressure of 0 to 40 MPa. The working pressure of the microscopic model holder is 0 to 5 MPa, as shown in Figure 3. The microscopic model imaging system comprises a microscope (Leica DM450, Leica Biosystems, Shanghai, China) and a computer. The dimethyldichlorosilane used in the experiment is produced by Aladdin, with a purity greater than 98.5%. The crude oil used is S crude oil, with an underground oil viscosity ranging from 24 to 452 cP; a viscosity of 86.5 cP is selected. The surface oil density is 0.971 t/m^3.

The experimental condition of microscopic model displacement is temperature 60 °C and pressure 15 Mpa, which is consistent with the actual reservoir temperature and pressure. Furthermore, the pore shape of the microscopic model is consistent with that of the actual core, which means the experimental results are similar to those of the actual reservoir.

The experimental steps are as follows:

① Age the model using dimethyldichlorosilane, allowing it to stand for 12 h to change the wettability of the model surface to oil wet.

② Utilize a saturated model with the S Oilfield crude oil and let it stand for 24 h.

③ Use formation water from the S Oilfield to inject water into the model until the produced fluid's water cut reaches 98%. Record the experimental process.

④ Inject polymer into the model at a rate of 0.05 mL/min until the produced fluid's water cut reaches 70%. Record the experimental process.

⑤ Inject the SP system into the model at a rate of 0.05 mL/min until the produced fluid's water cut reaches 98%. Record the experimental process.

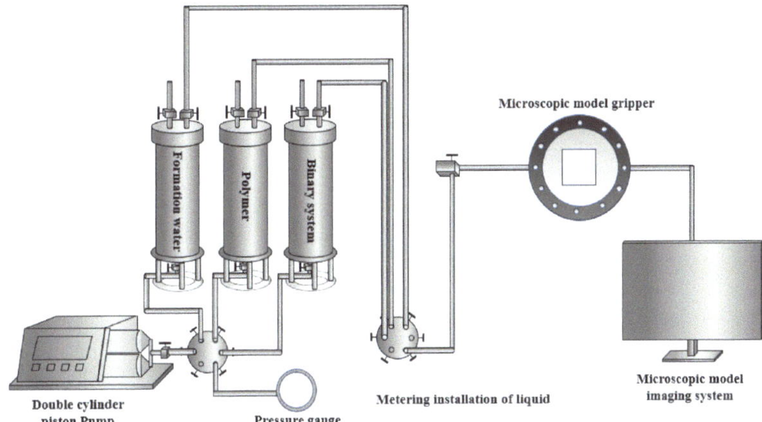

Figure 2. Microscopic model displacement experiment flowchart.

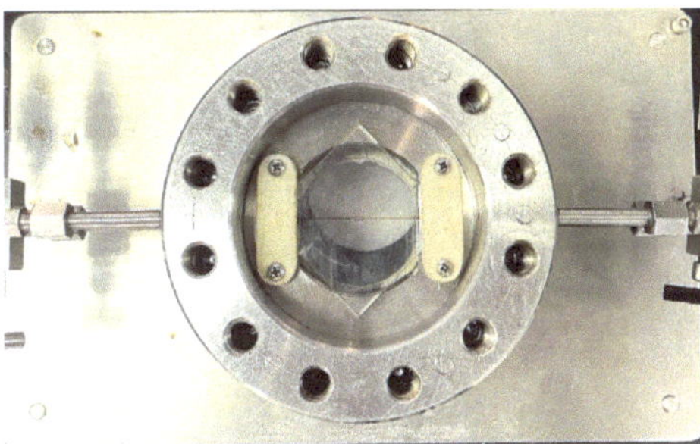

Figure 3. Microscopic model gripper.

2.3. Core Flooding Experiment

The experimental equipment used for core flooding is the Newmaze Analysis High-Temperature High-Pressure Displacement Nuclear Magnetic Resonance Testing System (MesoMR-HTHP, Suzhou Newmag Analytical Instrument Corporation, Suzhou, China). This system includes online nuclear magnetic resonance instruments, nuclear magnetic resonance analysis devices, a high-temperature high-pressure core holder, a high-temperature high-pressure displacement device, and connecting pipelines, as shown in Figure S1 of the Supplementary Materials. The pressure range of the high-temperature high-pressure core holder is 0 to 20 MPa, and the temperature range is 0 to 80 °C. The high-temperature high-pressure displacement device is of the MR-HTHP type, with a pressure range of 0 to 20 MPa and a temperature range of 0 to 80 °C. The formation water used in the experiment is simulated formation water containing Mn^{2+}, prepared from manganese chloride and deionized water. The T_2 signals were converted into oil saturation in pores through mathematical processing, as shown in Figure S2 of the Supplementary Materials.

The core oil displacement experiment steps are as follows:

(1) Begin by washing the core and measuring its length, diameter, permeability, and weight. Subsequently, evacuate the core, saturate it with simulated formation water, and scan the nuclear magnetic resonance T_2 spectrum upon completion.

(2) Employ simulated formation water containing Mn^{2+} for constant-rate displacement of the core, thoroughly replacing the simulated formation water to eliminate water signals within the core.

(3) Utilize crude oil at a constant rate of 0.5 mL/min for core displacement, continuing until the outlet of the core reaches 100% oil saturation. Establish the initial oil saturation and measure the nuclear magnetic resonance T_2 spectrum after oil saturation.

(4) Employ simulated formation water at a constant flow rate of 0.5 mL/min for water flooding, continuing until the outlet water cut reaches 70%. Monitor nuclear magnetic resonance signals in real-time during the experiment and record liquid and oil production at different time intervals.

(5) Conduct polymer flooding at a constant rate of 0.3 mL/min. Halt polymer injection when 0.4 pore volumes (PVs) have been injected, followed by continuous water flooding at 0.3 mL/min until the outlet water cut reaches 98%. Pause simulated formation water injection, continuously monitor nuclear magnetic resonance signals, and record liquid and oil production at different time intervals.

(6) Employ an SP system for constant-rate displacement at 0.3 mL/min. Halt the injection when 0.3 PV of the SP system has been introduced, followed by continuous water flooding at 0.3 mL/min until the outlet water cut reaches 98%. Continuously monitor nuclear magnetic resonance signals during the experiment and record liquid and oil production at different time intervals.

3. Results and Discussion

3.1. Mechanism of Enhanced oil Recovery by Surfactant–Polymer Flooding

3.1.1. Improve Displacement Efficiency

Surface tension experiments were carried out by varying the mass fractions of surfactants in contact with S crude oil, and the experimental results are presented in Figure 4. It is evident from the graph that the interfacial tension was significantly reduced by the surfactant solution, reaching values below 10^{-3} mN/m. Specifically, within the range of 0.30 wt% to 0.36 wt% of surfactant mass fraction, the interfacial tension varied between 0.25 and 0.8×10^{-3} mN/m.

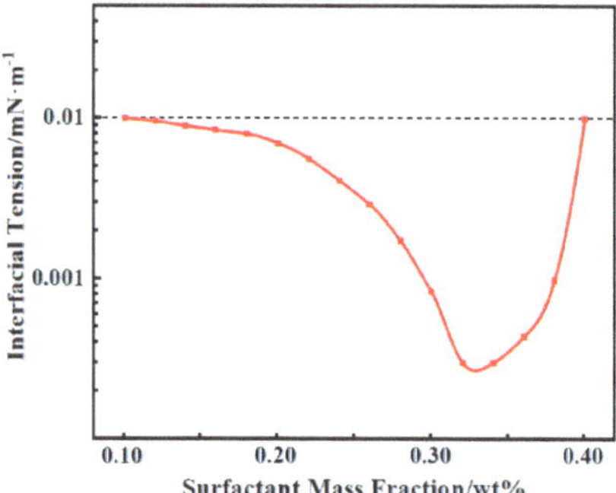

Figure 4. Interface tension curves under different mass fractions of surfactants.

In the SP system, a robust emulsification effect on crude oil is demonstrated by surfactants. Surfactant molecules can form large-sized micelles, partially encapsulating the crude oil in the aqueous solution of surfactants, leading to the creation of oil-in-water emulsions, as illustrated in Figure 5a. Emulsions are less prone to adsorption in the pores, thereby improving the flowability of crude oil in the pores, reducing residual oil, and enhancing oil recovery. The mass fractions of surfactant solutions achieving ultra-low interfacial tension were 0.30 wt%, 0.32 wt%, 0.34 wt%, and 0.36 wt%. The emulsion droplet size distribution and average size were measured.

As the concentration of surfactant increased, the average droplet sizes were 712.4 ± 200 nm, 1562.0 ± 200 nm, 955.4 ± 100 nm, and 1222.1 ± 100 nm, respectively.

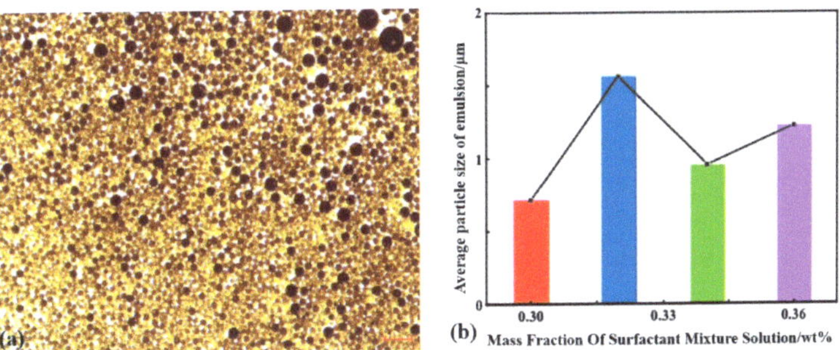

Figure 5. Microscopic image of oil in water system (**a**) and particle size distribution and average particle size of emulsion droplets (**b**).

The position and width of the peak in the emulsion droplet size distribution curve indicate the emulsification efficiency. A peak closer to the larger size and a wider width correspond to poorer emulsification, whereas a smaller droplet size and a narrower distribution indicate higher emulsion stability. From the peak widths of different surfactants, it can be observed that the emulsion formed with 0.34 wt% surfactant had the narrowest peak, indicating a higher level of droplet size uniformity. The emulsions formed with 0.32 wt% and 0.36 wt% surfactant follow, while the emulsion with 0.30 wt% surfactant exhibits lower droplet size uniformity, as shown in Figure 5b.

3.1.2. Extend the Swept Volume

The micro-scale displacement experiments in the S model revealed that water flooding was significantly influenced by the viscosity contrast between oil and water. Upon entering the channels, water induced severe fingering effects, resulting in poor displacement efficiency for the oil on both sides. Water flooding created high-permeability flow channels, and subsequent water injection progressed along these high-permeability channels. This led to a substantial amount of remaining oil in the unreached areas, as depicted in Figure 6a.

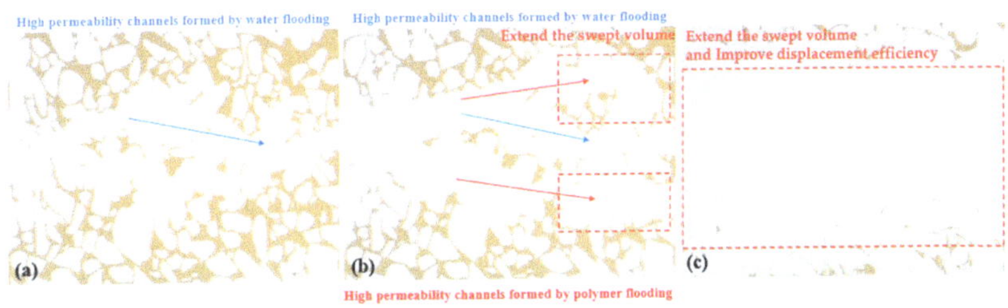

Figure 6. High-permeability channels formed by water flooding (**a**), polymer flooding forms high-permeability channels (**b**) and distribution of remaining oil after SP flooding (**c**).

Following the subsequent injection of polymers, the viscosity in the polymer allows for the control of the flow velocity in high-permeability zones, effectively managing the mobility and expanding the sweep volume. However, even after polymer flooding, significant residual oil remained in both the swept and unswept regions, with the distribution of residual oil in the swept region being more complex than after water flooding. Subsequent water flooding post polymer flooding encountered difficulties in displacing the remaining oil in the swept region, as illustrated in Figure 6b.

The SP system served to increase the viscosity of the aqueous phase, reducing its mobility and effectively improving the oil–water mobility ratio, thereby mitigating lateral fingering phenomena. Residual oil that could not be displaced during the polymer flooding stage experienced an increase in frictional forces with the SP system due to the reduced mobility ratio. The distribution of residual oil post SP flooding as the SP system began to flow is depicted in Figure 7. There existed mutual adsorption between the SP system and the pore walls, leading to significant frictional forces during displacement. This reduction in the flow velocity of the SP system in the pores diminished the velocity difference between high-permeability and low-permeability channels. The injection of the SP system adjusted the water imbibition profile, expanding the sweep volume and passively mobilizing the residual oil on both sides of the model, as illustrated in Figure 6c.

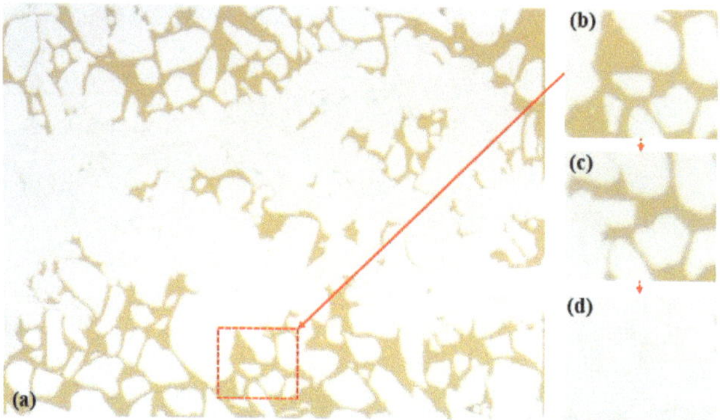

Figure 7. Remaining oil distribution after polymer flooding (**a**), part of the remaining oil (**b**), oil–water flow ratio decrease and driven remaining oil (**c**), completed remaining oil displacement (**d**).

The micro-scale displacement experiments in the S model revealed that, on the one hand, the polymer in the SP system effectively reduced the oil–water mobility ratio, thereby expanding the sweep volume coefficient. On the other hand, the surfactant in the SP system lowered the interfacial tension between oil and water, emulsifying the crude oil and consequently enhancing the overall oil recovery.

3.2. Potential to Improve Oil Displacement Efficiency

After conducting ultimate water flooding experiments using three viscosity grades of crude oil, namely 45.7 cP, 86.5 cP, and 291.1 cP, the total injected volume for ultimate water flooding of the three viscosity grades of crude oil was around 230 pore volumes (PVs). The ultimate water flooding efficiencies for the three viscosity grades were 75.23%, 73.22%, and 69.87%, with effective enhancements of 25.23%, 28.99%, and 33.67%, respectively, as illustrated in Figure 8.

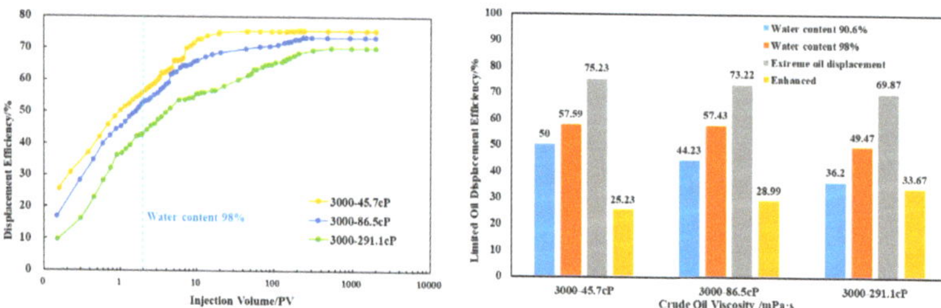

Figure 8. Water flooding experiments of the S Oilfield (3000 mD).

Selecting crude oil with a viscosity of 86.5 cP, ultimate polymer flooding and ultimate SP flooding experiments were conducted. In the case of the S Oilfield, the ultimate polymer flooding increased the recovery rate from 73.22% to 85.24%, showcasing a substantial potential for improved oil recovery. The total injected volume for ultimate polymer flooding was reduced from 230 PV in ultimate water flooding to 45 PV, marking a reduction of 185 PV and a reduction rate of 80.43%. For ultimate SP flooding, the total injected volume decreased from 230 PV to 3 PV, indicating a reduction of 227 PV and a reduction rate of 98.7%. The ultimate SP flooding achieved the maximum oil recovery efficiency more rapidly compared to polymer flooding.

Comparing the results of ultimate flooding experiments in the S Oilfield, it is evident that during the high water cut period in water flooding, both polymer and SP flooding can effectively enhance oil recovery. The EOR measure begins to be implemented at about 80% of water content. Polymer flooding reaches the displacement limit at 0.88 PV and 13.1 PV. SP flooding starts at 1.08 PV and reaches the displacement limit at 12.7 PV. The difference in ultimate oil recovery efficiency between polymer flooding and SP flooding is negligible. However, SP flooding significantly reduces the total injected PV from 45 PV to 3 PV, as depicted in Figure 9.

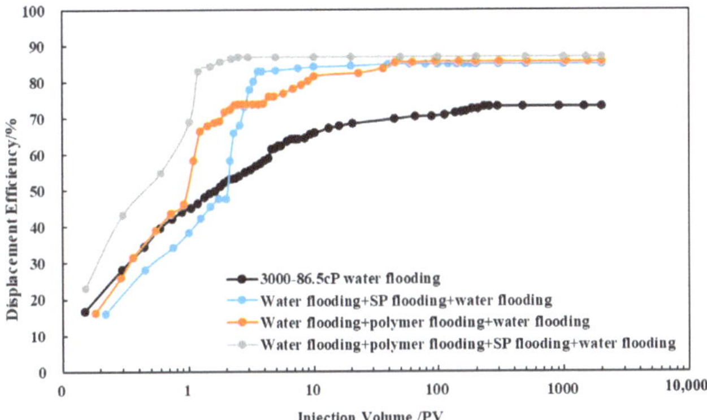

Figure 9. Comparison of displacement experiments (3000 mD).

After polymer flooding in the S Oilfield, the potential for improvement in ultimate oil recovery efficiency ranges from 10% to 14%. Measures such as SP flooding can effectively enhance the ultimate oil recovery efficiency, reducing the total injected PV required to approach the ultimate efficiency. This can significantly lower the time and cost associated with oil recovery.

3.3. Potential to Expand Swept Volume

Using numerical simulation methods and focusing on the Bohai S Oilfield, this study delved into the development potential of the oilfield, with a particular emphasis on key indicators such as target recovery rate and maximum sweep coefficient. Detailed data on the geology, reservoir properties, reservoir characteristics, and development methods of the S Oilfield were utilized for a comprehensive analysis. The parameters in Table 1 were designed to establish a high-permeability, heterogeneous conceptual model of the S Oilfield. The purpose of this model was to gain profound insights into the development prospects of the S Oilfield. Through predictions of key indicators like target recovery rate and sweep coefficient, it aimed to provide theoretical support for the oilfield's development.

Table 1. Conceptual Model Parameters for the S Oilfield.

Parameter	Taking Values	Parameter	Taking Values
Grid step(m)	500 (direction I); 250 (direction J); 48 (direction K)	Oil saturation	0.65
Number of grids	91 × 50 × 6	Oil viscosity (mPa·s)	86.5
Porosity (%)	32	Stratum water viscosity (mPa·s)	0.5
Permeability variation coefficient	0.35	Polymer injection amount (PV)	0.4
Average permeability(mD)	3419.67	SP injection amount (PV)	0.3

The foundational geological parameters are as follows: a five-spot pattern well arrangement, with well spacing of 175 m × 150 m and reservoir thickness of 48 m, vertically divided into three distinct permeability layers (from top to bottom, each with a thickness of 8 m and permeabilities of 4813 mD, 2987 mD, and 2459 mD, respectively). The model volume is set at 600×10^4 m^3.

Applying polymer flooding followed by SP flooding to the conceptual model of the S Oilfield under a five-spot pattern well arrangement allows for the determination of the target recovery rate and maximum sweep coefficient (as shown in Figure 10). The injection rate is set at the actual field well group daily injection rate of 6340 m^3. Utilizing data provided by the S Oilfield, a standard relative permeability curve is obtained (as shown in Figure 11). Reservoir engineering methods are then employed to calculate the maximum sweep coefficient and target recovery rate. This comprehensive assessment aims to evaluate the development potential of the S Oilfield.

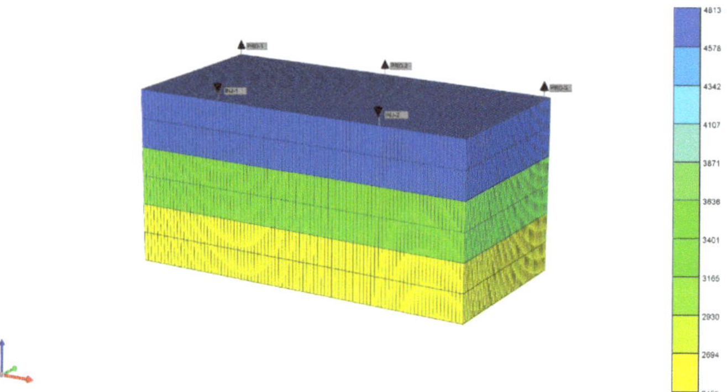

Figure 10. The conceptual model for the high-permeability heterogeneous S Oilfield.

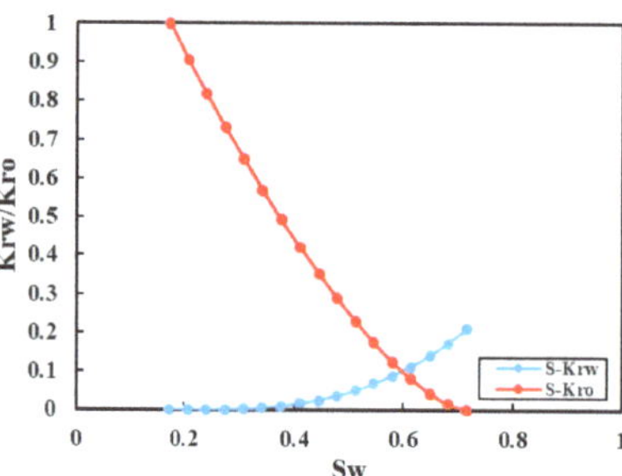

Figure 11. The permeability curve of conceptual model.

(1)   Evaluation of Ultimate Oil Recovery Efficiency Potential

Assuming a 100-year production period, an initial water flooding simulation was conducted. When the water cut reached 70%, a polymer with a concentration of 0.3 mg/L was injected at 0.4 PV. After the polymer injection, a second water flooding was initiated. When the water cut reached 98% in the second water flooding, an SP system with a concentration of 0.34 mg/L was injected at 0.3 PV. Subsequently, three consecutive water flooding cycles were simulated, as illustrated in Figure 12.

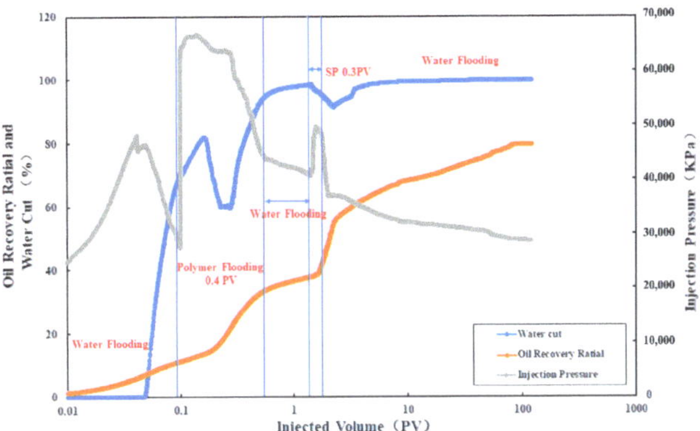

Figure 12. Changes in pressure, recovery rate and water content of SP flooding after polymer flooding.

The graph indicates that at the end of the first water flooding, the water cut was 70%, and the recovery rate was 11.34%. During the first water flooding, the pressure initially increased rapidly to 48.04 MPa, and as crude oil was gradually produced, the pressure rapidly decreased. When the cumulative injection volume reached 0.096 PV (end of the first water flooding), the pressure finally decreased to 27.49 MPa. The pressure decrease suggests that after a period of water flooding, the reservoir has formed high-permeability channels and water injection tends to flow through these high-permeability layers, influencing the crude oil in the high-permeability layers.

When the water cut reaches 70%, polymer injection is initiated. After polymer injection, a rapid pressure increase to 66.23 MPa is observed, maintaining a brief stability, while the water cut increases to 83.83%. This phenomenon is attributed to the significant viscosity of the polymer, leading to retention in the high-permeability layer due to chemical adsorption upon entering the porous reservoir. This retention increases flow resistance, resulting in a rapid increase in injection pressure. As a large amount of water is introduced into the reservoir through the first water flooding, the water cut does not immediately decrease after polymer injection. In reality, the water cut only decreases rapidly when the polymer passes through the advantageous channels formed by the sealed water flooding. It is noteworthy that at a cumulative injection volume of 0.157 PV, the pressure begins to slowly decrease, followed by a rapid decrease. Simultaneously, the recovery rate rapidly increases, and the water cut decreases rapidly to 60.14%. This indicates that the polymer has successfully temporarily plugged the major channels in the high-permeability layer (4813 mD), redirecting subsequent fluids to other layers, thus expanding the sweep and enhancing oil recovery, leading to a rapid increase in recovery rate. When the cumulative injection volume reaches 0.28 PV, the water cut begins to increase rapidly, and the increase in recovery rate slows down. At a cumulative injection volume of 0.494 PV (end of polymer injection), the water cut increases to 92.62%, the recovery rate increases to 32.71%, and the pressure decreases to 45.32 MPa. In the second water flooding stage, the rate of pressure reduction, as well as the rates of increase in water cut and recovery rate, slows down. Ultimately, at the end of the second water flooding, the cumulative injection volume reaches 1.39 PV, the pressure is 40.64 MPa, the water cut is 98.01%, and the recovery rate is 37.74%.

When the water cut reaches 98%, an SP system is injected at 0.3 PV, and it is observed that the pressure rapidly rises to 49.3 MPa, followed by a sharp decline. Simultaneously, there is a rapid decrease in water cut and a rapid increase in recovery rate. After the completion of the SP flooding, three consecutive water flooding cycles are conducted, and it is found that the recovery rate continues to increase rapidly, indicating the successful entry of polymer-carrying surfactant into more pore volumes (see Figure 12). This allows the washing effect of the surfactant to be effectively utilized, generating a synergistic effect. In this process, the main mechanisms include the temporary plugging effect of the polymer in the SP system and the alteration of rock wettability by the surfactant. By changing the wettability of the rock surface from oil wet to water wet, the third water flooding can still ensure a rapid increase in recovery rate. It is noteworthy that when the cumulative injection volume reaches 2.23 PV, the rate of increase in recovery rate begins to slow down, and the water cut gradually rises. When the water cut reaches 98% again, the recovery rate is 62.87%. Furthermore, during the third water flooding, the surfactant's ability to alter rock wettability remains effective, achieving continuous displacement of various oil layers and increasing crude oil production with a relatively long effective period. When the cumulative injection volume reaches 119.5 PV, the final recovery rate can reach 79.59%.

Through comparison, it is observed that at the extreme water cut of 98%, the recovery rate of polymer flooding followed by subsequent water flooding increases by 26.4%, while the recovery rate of SP system flooding followed by subsequent water flooding increases by 25.13%. Although the degrees of recovery rate improvement are similar between the two methods, the latter has a lower injection pressure, faster effectiveness, less injection volume, and shorter exploitation period, making it cost effective for field applications.

(2) Maximum Swept Efficiency

Based on the observations from Figure 13, after polymer flooding and six months of subsequent water flooding, there is still a significant amount of remaining oil in the swept area, indicated by the red and light-yellow regions. Calculations using reservoir engineering methods show a moderate sweep efficiency of 61.04% and a strong sweep efficiency of 37.11%. This indicates that after polymer flooding, the reservoir is still primarily characterized by moderate sweep efficiency.

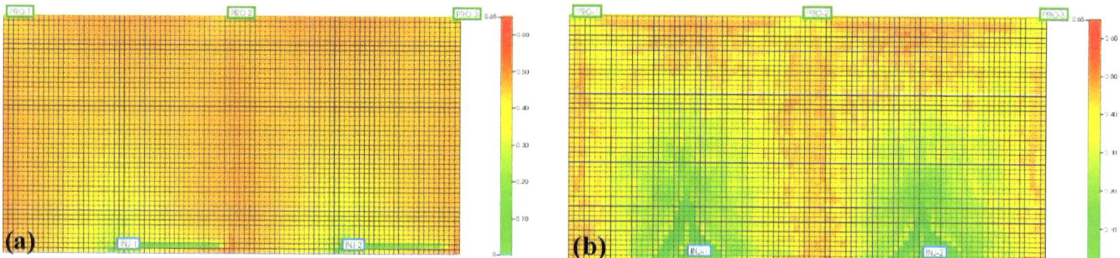

Figure 13. The change in sweep range after 6 months of follow-up of water flooding after polymer flooding (**a**) and 6 months of follow-up of water flooding after SP flooding (**b**).

In contrast, after the injection of the SP system and six months of three consecutive water flooding cycles, it is evident that the injection of the SP system has further enhanced its effectiveness in recovering remaining oil. The calculations reveal a moderate sweep efficiency of 34.66% and a strong sweep efficiency of 65.34% for the SP system. Compared to polymer flooding, the injection of the SP system has increased the strong sweep efficiency of the reservoir by 28.23%. This injection significantly expands the extent of reservoir sweeping, slightly reduces the dead oil zone areas on both sides, and greatly reduces the oil saturation. In summary, this indicates that the reservoir in the S Oilfield has the potential for long-term development, and the adoption of appropriate enhanced oil recovery techniques is likely to effectively promote the efficient development and utilization of the S Oilfield after polymer flooding.

The swept area under water flooding up to the extreme water cut is primarily characterized by moderate sweep efficiency (constituting 50% to 90%). Polymer flooding up to the limit helps increase the extent of sweeping, with some of the moderate swept areas transforming into strong swept areas (constituting 15% to 30%). The application of the SP system further enhances the strong swept regions.

3.4. Potential to Enhanced Oil Recovery

During online nuclear magnetic resonance (NMR) core flooding experiments for oil displacement, a significant reduction in the signal amplitude of the T_2 spectrum was observed as the water displacement pore volume (PV) increased, as shown in Figure 14a. Passive utilization was predominantly observed throughout the water flooding phase, with higher amplitudes. Notably, the medium and large pores exhibited the highest degrees of mobilization, accounting for 47.66% and 50.11%, respectively, as depicted in Figure 15a. Conversely, mobilization in small pores was relatively limited. In the water flooding stage, the medium and small pores contributed the most to oil displacement, with respective contribution rates of 59.30% and 20.78%. The contribution rate of large pores was 18.34%, while that of micro-pores was comparatively lower, resulting in a stage recovery rate of 43.25%, as illustrated in Figure 16. The distribution of remaining oil after water flooding was highest in the medium pores at 53.81%, followed by small pores at 23.40%. Large pores contained a smaller portion of saturated oil at 15.83%, and micro-pores exhibited the least saturation at 6.97%, as shown in Figure 17a. Following the conclusion of water flooding, the remaining oil was predominantly distributed in the medium and small pores, with medium pores being the primary reservoir, followed by small pores. Utilization in micro-pores remained relatively low, and the proportion of remaining oil was relatively unchanged, as indicated in Figure 14b.

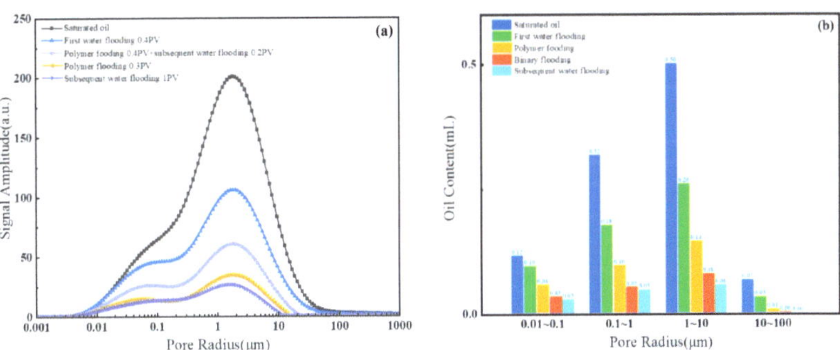

Figure 14. T_2 spectra for each stage (**a**) and oil content distribution in different pores at different stages (**b**).

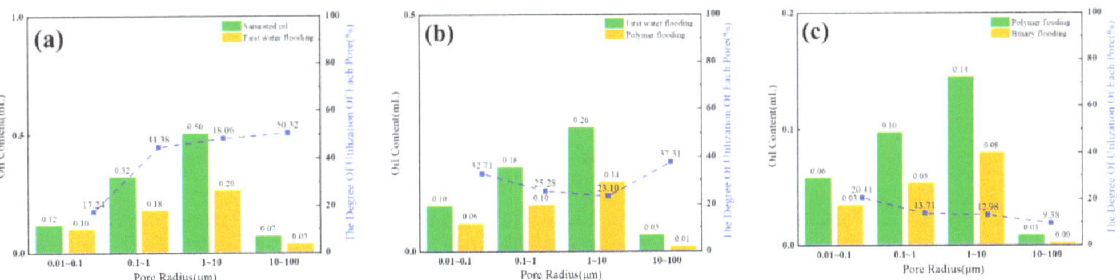

Figure 15. Distribution of oil content and effective use of pore development in water flooding stage (**a**), polymer flooding stage (**b**) and SP flooding stage (**c**).

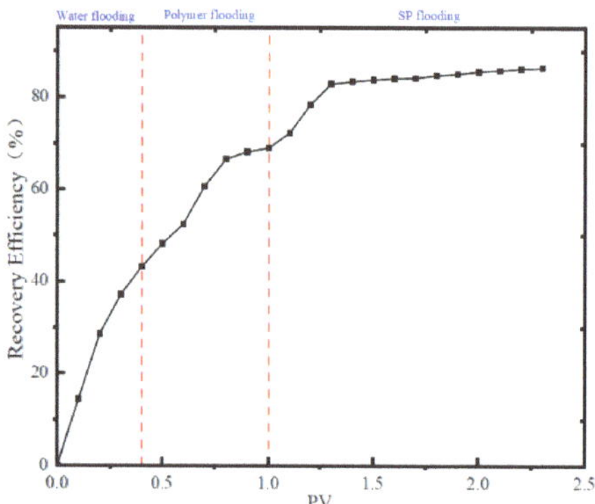

Figure 16. Recovery rates at different stages.

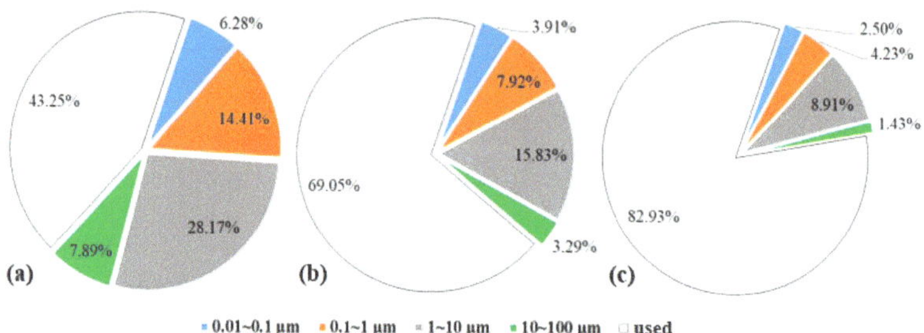

Figure 17. Distribution of remaining oil in water flooding stage (**a**), polymer flooding stage (**b**) and SP flooding stage (**c**).

With an increase in polymer flooding pore volume (PV), there was a significant reduction in the signal amplitude of the T_2 spectrum, as illustrated in Figure 14a. Throughout the polymer flooding phase, there was an overall higher passive utilization amplitude, with the highest degree of mobilization observed in micro-pores at 34.03%. The mobilization levels in other pore types were approximately around 25%, as depicted in Figure 15b. In the polymer flooding stage, the most substantial oil displacement occurred in small and medium pores, contributing rates of 25.15% and 47.81%, respectively. In comparison to water flooding, the utilization in micro-pores showed an increased contribution rate, resulting in a stage recovery rate of 25.81% and an overall recovery rate of 69.06%, as shown in Figure 16. The distribution of remaining oil after polymer flooding revealed that medium pores had the highest proportion at 28.17%, followed by small pores at 14.41%. Large pores exhibited a comparatively lower proportion of remaining oil at 7.89%, and micro-pores had the least remaining oil at only 6.28%, as depicted in Figure 17b. Following the conclusion of polymer flooding, the remaining oil was predominantly distributed in medium and small pores, with a certain degree of mobilization observed in micro-pores and a decline in remaining oil in large pores, as indicated in Figure 14b.

With an increase in SP flooding pore volume (PV), there was a significant reduction in the signal amplitude of the T_2 spectrum, as depicted in Figure 14a. Throughout the SP flooding phase, there was an overall higher passive utilization amplitude, with the highest degree of mobilization observed in micro-pores and small pores at 20.33% and 15.74%, respectively, as shown in Figure 15c. In the SP flooding stage, the contribution rates of micro-pores and small pores showed a noticeable increase, reaching 10.20% and 26.54%, respectively, resulting in a stage recovery rate of 17.39% and an overall recovery rate of 86.45%, as illustrated in Figure 16. The distribution of remaining oil after SP flooding indicated that medium pores had the highest proportion at 15.83%, followed by small pores at 7.92%. Large pores exhibited a comparatively lower proportion of remaining oil at 6.17%, and micro-pores had the least remaining oil at only 3.09%, as shown in Figure 17c. Following the conclusion of SP flooding, the remaining oil was predominantly distributed in medium pores with a relatively high proportion at 8.91%, while the proportions of remaining oil in small and micro-pores decreased, as indicated in Figure 14b.

4. Conclusions

Through the aforementioned experiments in high-permeability heterogeneous offshore oilfields, it was observed that, compared to polymer flooding, surfactant–polymer flooding not only expanded the swept volume but also enhanced oil recovery on the basis of polymer flooding. This is specifically manifested in the following aspects:

1. Microscopic displacement experiments revealed that, in the surfactant–polymer system, polymers effectively reduced the oil–water mobility ratio and increased the volumetric sweep efficiency coefficient, and concurrently, surfactants in the surfactant–polymer system reduced the interfacial tension of oil and water, emulsifying crude oil, thereby enhancing crude oil recovery.
2. Following polymer flooding, the potential for incremental oil recovery efficiency ranged from 10% to 14%. SP flooding effectively increases the incremental oil recovery efficiency, reduces the total injected PV required to approach the incremental efficiency, and significantly lowers the cost of oil recovery time.
3. Driving to the ultimate limit in polymer flooding contributes to an increase in the extent of sweep, with a portion of the medium sweep region transforming into a strong sweep region (constituting 15% to 30%). SP flooding can further elevate the extent of the strong sweep region.
4. In situ nuclear magnetic resonance core flooding experiments verified the tremendous potential of SP flooding in enhancing oil recovery. Taking the S Oilfield as an example, SP flooding can increase recovery by 13.83% on the basis of polymer flooding.

Supplementary Materials: The following supporting information can be downloaded at: https://www.mdpi.com/article/10.3390/polym16142004/s1, Table S1: Comparison of Actual Reservoir Conditions in S Oilfield with Laboratory Experimental Conditions; Figure S1: Newmaze Analysis High-Temperature High-Pressure Displacement Nuclear Magnetic Resonance Testing System; Figure S2: The conversion result of T_2 signal.

Author Contributions: Conceptualization, Y.L. and J.H.; methodology, L.G. and J.H.; validation, K.M. and X.C.; formal analysis, X.C.; investigation, Z.Z. and J.H.; data curation, K.M.; writing—original draft preparation, Y.L.; writing—review and editing, Y.L. and L.G.; visualization, K.M.; project administration, X.C. and Z.Z.; funding acquisition, Y.L. All authors have read and agreed to the published version of the manuscript.

Funding: The authors gratefully appreciate the financial support of the National Natural Science Foundation of China (Grant No. 52174046).

Institutional Review Board Statement: Informed consent was obtained from all subjects involved in the study.

Data Availability Statement: The original contributions presented in the study are included in the article/Supplementary Material, further inquiries can be directed to the corresponding author.

Conflicts of Interest: Yingxian Liu, Lizhen Ge, Kuiqian Ma, Xiaoming Chen and Zhiqiang Zhu are employed by CNOOC China Limited Tianjin Branch. The remaining author declares that the research was conducted in the absence of any commercial or financial relationships that could be construed as potential conflicts of interest.

References

1. Silva, I.G.; Melo, D.; Aparecida, M.; Luvizotto, J.M.; Lucas, E.F. Polymer flooding: A sustainable enhanced oil recovery in the current scenario. In Proceedings of the Latin American and Caribbean Petroleum Engineering Conference, Buenos Aires, Argentina, 15–18 April 2007.
2. Samanta, A.; Bera, A.; Ojha, K.; Mandal, A. Comparative studies on enhanced oil recovery by alkali–surfactant and polymer flooding. *J. Pet. Explor. Prod. Technol.* **2012**, *2*, 67–74. [CrossRef]
3. Han, D.-K.; Yang, C.-Z.; Zhang, Z.-Q.; Lou, Z.-H.; Chang, Y.-I. Recent development of enhanced oil recovery in China. *J. Pet. Sci Eng.* **1999**, *22*, 181–188. [CrossRef]
4. Abidin, A.Z.; Puspasari, T.; Nugroho, W.A. Polymers for enhanced oil recovery technology. *Procedia Chem.* **2012**, *4*, 11–16. [CrossRef]
5. Pope, G.A. Overview of chemical EOR | enhanced oil recovery, surfactant. In *Casper EOR Workshop*; Center for Petroleum and Geosystems Engineering, The University of Texas at Austin: Austin, TX, USA, 2007.
6. Liu, L.; Zhao, M.; Pi, Y.; Fan, X.; Cheng, G.; Jiang, L. Experimental Study on Enhanced Oil Recovery of the Heterogeneous System after Polymer Flooding. *Processes* **2023**, *11*, 2865. [CrossRef]
7. Luo, F.-Q.; Gu, X.; Geng, W.-S.; Hou, J.; Gai, C.-C. Study on Screening Criteria of Gel-Assisted Polymer and Surfactant Binary Combination Flooding after Water Flooding in Strong Edge Water Reservoirs: A Case of Jidong Oilfield. *Gels* **2022**, *8*, 436. [CrossRef] [PubMed]
8. Pu, H.; Xu, Q. An update and perspective on field-scale chemical floods in Daqing oilfield, China. Paper SPE-118746-MS. In Proceedings of the SPE Middle East Oil and Gas Show and Conference, Manama, Bahrain, 15–18 March 2009.
9. Liu, H.; Li, J.L.; Yan, J.D.; Wang, W.J.; Zhang, Y.C.; Zhao, L.Q. Successful Practices and Development of Polymer Flooding in Daqing Oilfield, SPE123975. In Proceedings of the Asia Pacific Oil and Gas Conference and Exhibition, Jakarta, Indonesia, 4–6 August 2010.
10. Wang, F.L.; Li, X.; Li, S.Y.; Han, P.H.; Guan, W.T.; Yue, Y.H. Performance analysis and field application result of polymer flooding in low permeability reservoirs in Daqing Oilfield, SPE136904. In Proceedings of the Canadian Unconventional Resources and International Petroleum Conference, Calgary, AL, Canada, 19–21 October 2010.
11. Chen, X.; Feng, Q.; Liu, W.; Sepehrnoori, K. Modeling preformed particle gel surfactant combined flooding for enhanced oil recovery after polymer flooding. *Fuel* **2017**, *194*, 42–49. [CrossRef]
12. Han, M.; Xiang, W.T.; Zhang, J.; Jiang, W.; Sun, F.J. Application of EOR technology by means of polymer flooding in Bohai Oil Fields. In Proceedings of the International Oil & Gas Conference and Exhibition in China, Beijing, China, 5–7 December 2006; p. 104432.
13. Cheng, J.C.; Wei, J.G.; Song, K.P.; Han, P.H. Study on Remaining Oil Distribution after Polymer Injection, SPE133808. In Proceedings of the SPE Annual Technical Conference and Exhibition, Florence, Italy, 20–22 September 2010.
14. Ezell, R.G.; McCormick, C.L. Electrolyte- and pH-responsive polyampholytes with potential as viscosity-control agents in enhanced petroleum recovery. *J. Appl. Polym. Sci.* **2007**, *104*, 2812–2821. [CrossRef]
15. Rashidi, M.; Blokhus, A.M.; Skauge, A. Viscosity study of salt tolerant polymers. *J. Appl. Polym. Sci.* **2010**, *117*, 1551–1557. [CrossRef]
16. Sorbie, K.S. *Polymer Improved Oil Recovery*; Blackie and Son Ltd.: London, UK, 1991.
17. Xin, X.; Yu, G.; Chen, Z.; Wu, K.; Dong, X.; Zhu, Z. Effect of polymer degradation on polymer flooding in heterogeneous reservoirs. *Polymers* **2018**, *10*, 857. [CrossRef] [PubMed]
18. Delamaide, E.; Zaitoun, A.; Renard, G.; Tabary, R. Pelican Lake Field: First Successful Application of Polymer Flooding in a Heavy-Oil Reservoir. *SPE Res. Eval. Eng.* **2014**, *17*, 340–354. [CrossRef]
19. Liu, H.; Wang, Y.; Liu, Y.-Z. Mixing and injection techniques of polymer solution. In *Enhanced Oil Recovery–Polymer Flooding*, 1st ed.; Shen, P.-P., Liu, Y.-Z., Liu, H.-R., Eds.; Petroleum Industry Press: Beijing, China, 2006.
20. Chen, Z. Study on potential and distribution of remaining oil after polymer flooding in XX Oilfield. *IOP Conf. Ser. Earth Environ. Sci.* **2021**, *651*, 032070. [CrossRef]
21. Wang, D. Chapter 4—Polymer Flooding Practice in Daqing. In *Enhanced Oil Recovery Field Case Studies*; Sheng, J.J., Ed.; Gulf Professional Publishing: Houston, TX, USA, 2013; pp. 83–116.
22. Pogaku, R.; Mohd Fuat, N.H.; Sakar, S.; Cha, Z.W.; Musa, N.; Awang Tajudin, D.N.; Morris, L.O. Polymer flooding and its combinations with other chemical injection methods in enhanced oil recovery. *Polym. Bull.* **2018**, *75*, 1753–1774. [CrossRef]
23. Shuang, L.; Shaoquan, H.; Jie, L.; Guorui, X.; Bo, Z.; Yudong, Z.; Jianye, L. Study on EOR method in offshore oilfield: Combination of polymer microspheres flooding and nitrogen foam flooding. *J. Pet. Sci. Eng.* **2019**, *178*, 629–639.
24. Lizhen GXiaoming, C.; Gang, W.; Guohao, Z.; Jinyi, L.; Yang, L.; Mingxing, B. Analysis of the Distribution Pattern of Remaining Oil and Development Potential after Weak Gel Flooding in the Offshore LD Oilfield. *Gels* **2024**, *4*, 236.
25. Xie, X.; Feng, G.; Liu, L.; Li, Y.; Yi, R. Enhanced Oil Recovery Technology After Polymer Flooding in Offshore Oilfields. *Oil Gas Geol. Recovery* **2015**, *22*, 93–97.
26. Dong, H.; Fang, S.; Wang, D.; Wang, J.; Liu, Z.; Hou, W. Review of practical experience & management by polymer flooding at Daqing. Paper SPE-114342-MS. In Proceedings of the SPE Symposium on Improved Oil Recovery, Tulsa, OK, USA, 20–23 April 2008.
27. Hailong, C.; Zhaomin, L.; Fei, W.; Songyan, L. Investigation on in Situ Foam Technology for Enhanced Oil Recovery in Offshore Oilfield. *Energy Fuels* **2019**, *33*, 12308–12318.

28. Sun, L.; Wu, X.; Zhou, W.; Li, X.; Han, P. Enhanced Oil Recovery Technology with Chemical Flooding in Daqing Oilfield. *Pet. Explor. Dev.* **2018**, *45*, 636–645. [CrossRef]
29. Sheng, J.J. Critical review of alkaline-polymer flooding. *J. Pet. Explor. Prod. Technol.* **2017**, *7*, 147–153. [CrossRef]
30. Zhu, Y.; Ming, L. Studies on Ssurfactant–polymer Combination Flooding Formulations for a High Salinity Reservoir. In Proceedings of the SPE EOR Conference at Oil and Gas West Asia, Muscat, Oman, 22 March 2016.

Article

Designing Sustainable Hydrophilic Interfaces via Feature Selection from Molecular Descriptors and Time-Domain Nuclear Magnetic Resonance Relaxation Curves

Masayuki Okada [1,2]**, Yoshifumi Amamoto** [2,3]**and Jun Kikuchi** [1,2,4,*]

1. Graduate School of Bioagricultural Sciences, Nagoya University, 1 Furo-cho, Chikusa-ku, Nagoya 464-8601, Japan; masayuki.okada@a.riken.jp
2. RIKEN Center for Sustainable Resource Science, 1-7-22 Suehiro-cho, Tsurumi-ku, Yokohama 230-0045, Japan; yoshifumi.amamoto@a.riken.jp
3. Graduate School of Social Data Science, Hitotsubashi University, 2-1 Naka, Kunitachi 186-8601, Japan
4. Graduate School of Medical Life Science, Yokohama City University, 1-7-29 Suehiro-cho, Tsurumi-ku, Yokohama 230-0045, Japan
* Correspondence: jun.kikuchi@riken.jp; Tel.: +81-45-508-7221

Abstract: Surface modification using hydrophilic polymer coatings is a sustainable approach for preventing membrane clogging due to foulant adhesion to water treatment membranes and reducing membrane-replacement frequency. Typically, both molecular descriptors and time-domain nuclear magnetic resonance (TD-NMR) data, which reveal physicochemical properties and polymer-chain dynamics, respectively, are required to predict the properties and understand the mechanisms of hydrophilic polymer coatings. However, studies on the selection of essential components from high-dimensional data and their application to the prediction of surface properties are scarce. Therefore, we developed a method for selecting features from combined high-dimensional molecular descriptors and TD-NMR data. The molecular descriptors of the monomers present in polyethylene terephthalate films were calculated using RDKit, an open-source chemoinformatics toolkit, and TD-NMR spectroscopy was performed over a wide time range using five-pulse sequences to investigate the mobility of the polymer chains. The model that analyzed the data using the random forest algorithm, after reducing the features using gradient boosting machine-based recursive feature elimination, achieved the highest prediction accuracy. The proposed method enables the extraction of important elements from both descriptors of surface properties and can contribute to the development of new sustainable materials and material-specific informatics methodologies encompassing multiple information modalities.

Keywords: hydrophilic coating materials; time-domain nuclear magnetic resonance; contact angle; chemoinformatics descriptors; machine learning

Citation: Okada, M.; Amamoto, Y.; Kikuchi, J. Designing Sustainable Hydrophilic Interfaces via Feature Selection from Molecular Descriptors and Time-Domain Nuclear Magnetic Resonance Relaxation Curves. *Polymers* **2024**, *16*, 824. https://doi.org/10.3390/polym16060824

Academic Editors: Laurence Noirez, Japan Trivedi, Zheyu Liu, Xin Chen, Jianbin Liu and Zhe Li

Received: 31 January 2024
Revised: 27 February 2024
Accepted: 11 March 2024
Published: 15 March 2024

1. Introduction

The bioeconomy [1] and circular economy [2] are the keys to realizing a sustainable society. With the shift in focus toward the management of the life cycle of plastics [3], understanding the interfaces of materials has become crucial. For example, understanding the mechanisms underlying biological and chemical reactions, including microorganism reactions that degrade polyethylene, polystyrene, and polypropylene [4,5], enzyme reactions that degrade polyethylene terephthalate (PET) [6], and marine biofouling, which occurs at the interfaces of materials, is crucial [7]. Therefore, methods based on wettability, which indicates the hydrophilicity and hydrophobicity of a material, antifoulant release, self-renewability, temperature and pH changes, and biomimetics have been developed [8]. Biocompatible 2-methacryloyloxyethyl phosphorylcholine, developed by introducing phosphatidycholine, which is a component present in biological membranes [9] has a hydration layer formed on the polymer imparts strong antifouling properties [10]. A hydration layer, namely an intermediate water layer, formed on poly(2-methoxyethyl acrylate) (PMEA) plays a critical role in preventing fouling [11,12]. Thus, polymer hydrophilicity as well as hydrophobic coatings play crucial roles in controlling fouling. Superhydrophobic polymers exhibit antifouling properties owing to their low surface free energies [13]. In addition, elastomers based on silicone or polydimethylsiloxane are used to prevent fouling. However, the adhesion between the coating material and substrates is weak, and thus, various studies have been conducted to improve the adhesion using nanofiller mixtures [14].

The functionality of polymers are influenced by their microscopic molecular structures as well as their intricate molecular dynamics, including the behavior of polymer chains and their entanglement. Hence, comprehending and managing molecular dynamics is pivotal for effectively understanding and harnessing polymer properties [15]. The surface properties of materials, such as rigidity, affect their hydrophilicity [16]. Nuclear magnetic resonance (NMR) is a powerful tool for analyzing molecular dynamics, and the corresponding signals can be measured over a wide range of timescales, from picoseconds to milliseconds (ms) [17]. Thus, NMR spectroscopy can be employed to study the relationships between the functions and physical properties of materials and their structures [18]. Proton and carbon-13 nuclear magnetic resonance (^{1}H NMR and ^{13}C NMR) spectroscopies are frequently used in polymer development [18,19], and time-domain NMR (TD-NMR), a highly promising analytical tool, is extensively utilized to explore the impact of internal and external factors on the structure and properties of various materials, including polymers, fresh foods, processed food products, and agricultural items [20].

Because NMR spectroscopy generates a large amount of high-dimensional data pertaining to molecular dynamics, various measurement informatics technologies have been simultaneously developed to streamline the associated measurement process [21]. To optimize machine learning (ML) performance, the extraction of optimal features from raw data is crucial. Moreover, high-dimensional datasets often lead to overfitting issues [22]. To address these challenges, approaches involving dimensionality reduction and/or feature selection are employed. Methods such as principal component analysis [23], multidimensional scaling [24], and linear discriminant analysis [25] are used for reducing feature space dimensions owing to their high efficacy in identifying highly relevant descriptors, which are commonly referred to as key features and are particularly beneficial for ML applications [26]. Additionally, non-negative matrix factorization (NMF), partial least squares [27], and semi-supervised NMF have been used as dimension reduction methods [28], in which genetic algorithms [29,30] are utilized to meaningfully reduce the relaxation component to 10% [31]. Recursive feature elimination (RFE), a type of feature selection, is applied to perform quality control of polylactic acid processing [32] and antibacterial peptide development [33]. Recently, a new RFE approach that evaluates the "feature (variable) importance" based on support vector machine (SVM), random forest (RF), and gradient boosting machine (GBM) models as well as selects and eliminates the least important features has been proposed [34].

As mentioned before, analyzing microscopic molecular structures is also necessary for assessing the surface properties of materials. For instance, in an antifouling membrane [8], after the initial formation of conditioning films, microorganism adhesion occurs [35]. The antifouling ability of PMEA originates from the interactions between the constituent carboxy and methoxy groups with water molecules [36]. Hence, controlling the intermolecular interactions is crucial. Materials informatics (MI) involves the analysis of microscopic surficial molecular structures using informatics technology. To date, various MI models based on molecular descriptors have been developed using open-source tools, such as RDKit, which is widely utilized in chemoinformatics. These models aid in devising synthesis strategies for molecules, including inorganic nickel (II) salts, organic photosensitizers [37], and amphiphilic copolymers [38]. Although MI research is primarily focused on extracting essential physicochemical components, progress in integrating these findings with molecular dynamics has been limited. Specifically, studies on the application of diverse types of RFE algorithms to NMR transition curves are scarce.

In the present study, we constructed an ML model that incorporates both molecular and dynamics descriptors to predict the hydrophilicity of hydrophilic polymer coating materials. The conceptual framework of the study is illustrated in Figure 1. We conducted RF classification using a combination of RDKit descriptors, five distinct pulse sequences from TD-NMR spectroscopy, and different ultraviolet (UV) wavelengths applied during the manufacturing of the coating material. This study was focused on enhancing interpretability through the application of RFE as a feature selection method.

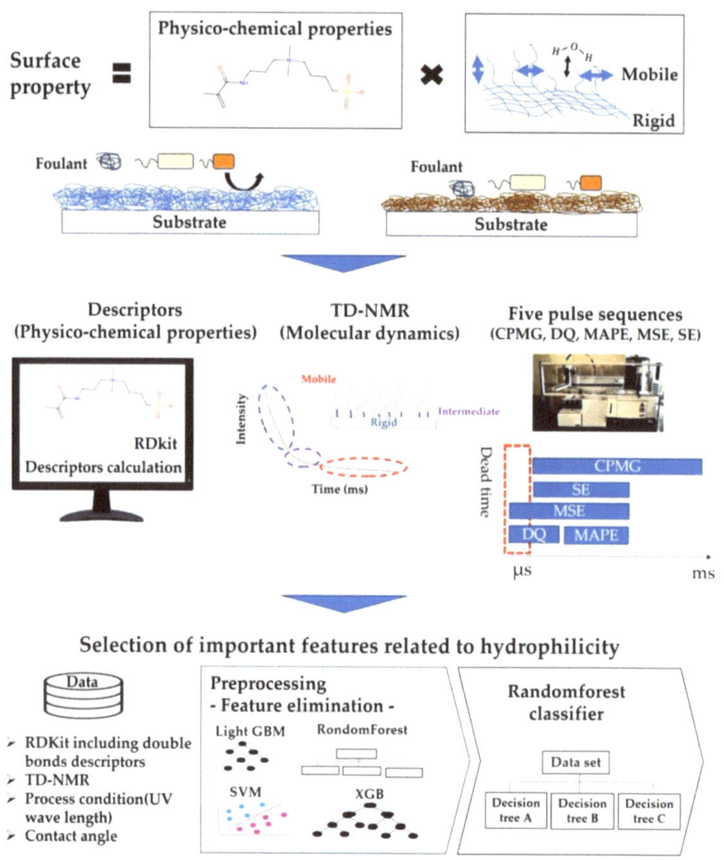

Figure 1. Conceptual framework of the study, illustrating the pivotal role of molecular descriptors and molecular dynamics in hydrophilicity development. The study encompassed calculations and measurements of these elements. The dataset, comprising RDKit, TD-NMR, contact angle data, and manufacturing process details, underwent feature selection via RFE. Preprocessed data were then utilized by RF classifier to identify crucial factors for building predictive models of hydrophilicity and investigating the underlying principles governing this trait.

2. Materials and Methods

2.1. Sample Preparation

2.1.1. Materials

N,N'-{[(2-acrylamide-2-[(3-acrylamidopropoxy) methyl] propane-1,3-diyl) bis (oxy)] bis(propane-1,3-diyl)}diacrylamide (FOM-3006), N,N',N''-triacryloydiethylenetriamine (FOM-3007), N,N'-diacryloyl-4,7,10-trioxa-1,13-tridecanediamine (FOM-3008), N,N',N'',N'''-

tetraacryloyltriethylenetetramine (FOM-3009), and [3-(Methacryloylamino)propyl]dimethyl (3-sulfobutyl)ammonium hydroxide inner salt (FOM-3010), *N-tert*-butylacrylamide (NTBA) were used as hydrophilic monomers (Figure S1). 2-Hydroxy-4′-(2-hydroxyethoxy)-2-methylpropiophenone (Irgacure 2959) was used as the photoinitiator. Methanol was used as the solvent to dissolve the monomers. FOM-3006, FOM-3007, FOM-3008, FOM-3009, FOM-3010, and methanol were obtained from Fujifilm Wako Pure Chemical Corporation (Osaka, Japan). NTBA and Irgacure 2959 were purchased from Tokyo Chemical Industry Co., Ltd. (Tokyo, Japan) and Sigma-Aldrich (Tokyo, Japan), respectively.

2.1.2. Surface Coating

Surface coatings were prepared by copolymerizing two types of hydrophilic monomers on PET sheets, which were adopted owing to the strong affinity of PET with hydrophilic polymers as well as a high recycling rate (approximately 85% in Japan) of PET [39]; owing to its high recycling rate, PET is one of the most sustainable plastics.

Two monomers were randomly selected and dissolved in methanol, which contained the photoinitiators listed in Table S1. Approximately 200 μL of each mixture was coated onto a PET sheet, with dimensions of approximately 3.5 cm × 3.5 cm (Cosmoshine A4360, TOYOBO, Osaka, Japan), using a micropipette. The mixing ratio of each monomer is listed in Table S1. The coated sheets were dried at 50 °C for 10 min using a constant temperature thermostatic dryer natural oven (NDO-420, TOKYO RIKAKIKAI CO., LTD., Tokyo, Japan). Subsequently, the sheets were exposed to UV light, with a wavelength of either 254 nm or 365 nm, inside a box using a handy UV light (SLUV-4, AS ONE CORPORATION, Osaka, Japan) for curing.

2.2. TD-NMR Measurements

TD-NMR measurements were conducted at 298 K using the Minispec mq20 NMR spectrometer (Bruker, Billerica, MA, USA) to assess the dynamics of the polymer chains within the hydrophilic coating. This equipment is equipped with Carr-Purcell Meiboom-Gill (CPMG) [28], double quantum (DQ) filter [40], magic sandwich echo (MSE) [40], solid echo (SE) [41], and magic and polarization echo (MAPE) [40]. The PET sheets were cut into square shapes measuring approximately 1 mm × 10 mm and placed in measurement tubes without any solvent. T_2 relaxation curves were obtained using five pulse sequences, viz, CPMG, DQ, MSE, SE, and MAPE. Because CPMG was a pulse sequence that could measure long relaxation times in the order of ms, information on rapid molecular mobility was obtained. SE could measure short relaxation times in the order of μs and was thus used to measure high-order structures, such as crystalline and amorphous structures. However, the application of SE was limited by its associated dead time. Thus, MSE, DQ, and MAPE pulse sequences, which could overcome this dead time issue, were employed. MSE consisted of DQ and MAPE, and DQ could measure extremely short relaxation times, whereas MAPE could measure relaxation times longer than those measured by DQ. The relaxation curves exhibited distinct time regions for the slow and fast mobile components of the polymers. Specifically, the MAPE, DQ, MSE, and SE sequences depicted the slow mobile components, while CPMG captured the fast mobile components in ms. MSE represented both slow and relatively fast mobile components. Furthermore, the DQ and SE sequences detected the slow mobile components with short relaxation times.

2.3. Contact Angle Measurement

Contact angle measurements were conducted using a contact angle meter (DMs-401, Kyowa Interface Science Co., Ltd., Saitama, Japan). A 2 μL droplet of clean water was dispensed using a microsyringe, and its side image was captured using the accompanying digital camera. From the obtained image, the contact angle was automatically calculated using the $\theta/2$ method. The contact angle of each specimen was measured thrice, and their average value was adopted as the final measured contact angle. Based on the measured contact angles, the films were divided into two groups: films with contact angles less than

25°, 30°, 35°, or 40° were categorized as 0, while those with contact angles exceeding 25°, 30°, 35°, or 40° were classified as 1. Such a classification was performed to avoid large differences in the amount of data after classification.

2.4. Generation of Molecular Descriptors

Molecular descriptors of the monomers were generated using a simplified molecular input line entry system [42], which transformed chemical structures into text representations, and RDKit (version 2023.3.2) (Table S2). The descriptors of all the monomers were selected (Table S3). The molecular descriptors of the copolymers were calculated based on the mixing ratio of the monomers and photoinitiators. Additionally, the number of chemical bonds, including double bonds (C–C, C=C, C–N, C–O), in each monomer was utilized as a descriptor. Furthermore, the bond distances between the vinyl groups were manually calculated based on individual bond numbers and bond lengths (C–C: 1.54 Å, C=C: 1.34 Å, C–N: 1.43 Å, and C–O: 1.43 Å).

2.5. Data Analysis

The data were processed using Python (version 3.10.12), scikit-learn library (version 1.2.2), LightGBM (version 1.2.2), and XGBoost (version 2.02). Autoscaling (standardization) was conducted for both the molecular descriptors and TD-NMR relaxation curves. Because the mobile molecules were assumed to be contributors to hydrophilicity, TD-NMR and data from five pulse sequences were employed in the analysis. Feature selection was executed using GBM-RFE, RF-RFE, SVM-RFE, and XGM-RFE. To ensure that the number of features remains less than the number of samples (=57), the number of features after reduction was set to 30. The features obtained via RFE were employed as explanatory variables, while the binary classification values of the contact angle were employed as target variables. RF classifiers were used to construct classification models. The data were split into training and test datasets using the holdout method. Hyperparameters were determined using cross-validation methods by applying GridSearchCV to the training data. Prediction accuracies were assessed based on the parameters: accuracy, precision, recall, and F1-score. The flow of data analysis is depicted in Figure S2, and the hyperparameter RF is shown in Table S4.

3. Results
3.1. Surface Coating

Surface coatings were applied by the photoinitiated copolymerization of acrylamide monomers on PET films. Both ionic (FOM-3010)) and nonionic (NBTA) monomers were utilized to alter the surface properties. Cross-linkers with varying numbers of vinyl groups were employed to stabilize the coating and regulate film dynamics. Upon exposure to UV light, the initially flowable liquid transformed into a cured solid with high viscosity. As a result, polyacrylamide was coated onto the PET films through the copolymerization of the monomers.

3.2. TD-NMR

Changes in chain dynamics due to surface modifications were assessed through TD-NMR measurements. Figure S3 displays the TD-NMR relaxation curves, which are correlated with the surface modification conditions, acquired for various pulse sequences. Noticeable distinctions in the relaxation curves were evident for CPMG, MSE, and SE sequences. Specifically, the CPMG and MAPE [40] appeared suitable for mobile components, suggesting their effectiveness in detecting components characterized by long relaxation times on the surface.

The impact of the presence or absence of cross-linkers on the relaxation curves was confirmed, as depicted in Figure 2. The evaluation of samples coated with NTBA using the CPMG sequence revealed a gradual attenuation in the intensity of T_2 relaxation. Conversely, the relaxation curve obtained using the MSE sequences exhibited a low intensity with minimal changes. However, for samples with cross-linkers, the CPMG relaxation curves exhibited minimal alterations, with sharper relaxations observed in the MSE sequences. These trends were accentuated with an increase in the number of vinyl groups. Polyacrylamide prepared with NTBA featuring a single vinyl group exhibited linear polymer chains, while FOM-3006, FOM-3007, FOM-3008, and FOM-3009, which possess multiple vinyl groups, displayed cross-linked or networked structures with reduced mobility. Thus, these findings underscore a disparity in the TD-NMR relaxation curves, stemming from the chain mobilities of polyacrylamide on the surface.

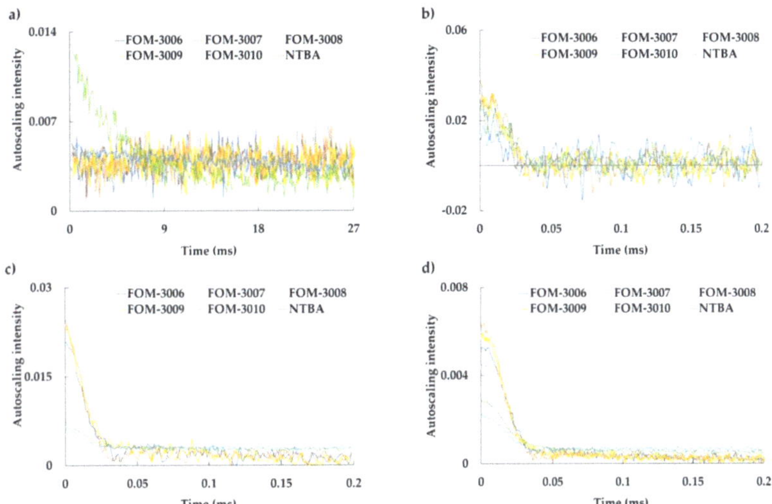

Figure 2. T_2 relaxation curves of each single-monomer polymer. (**a**) CPMG, (**b**) DQ, (**c**) MSE, and (**d**) SE. The blue, orange, gray, yellow, and light and dark green lines represent FOM-3006, FOM-3007, FOM-3008, FOM-3009, FOM-3010, and NTBA, respectively.

3.3. Contact Angle

The properties of the modified surfaces were evaluated through contact angle measurements. Figure 3 illustrates a histogram of the contact angles of the sample films. The contact angles of the modified surfaces exhibited a broad range of values, spanning from 5° to 80°. To determine the standard for binary classification, data analysis was performed according to the approach indicated in Section 2.5, and the best results were obtained at 40°. Therefore, 40° was set as the criteria for binary classification. The performance data for angles of 25°, 30°, and 35° are shown in Figure S4.

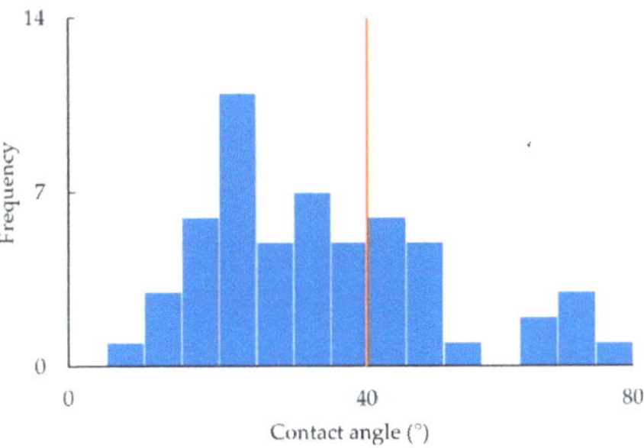

Figure 3. Contact angle histogram, with each bin representing a 5° interval ranging from 10° to 80°. The orange line represents 40°, which is the criteria for binary classification.

3.4. RFE

Feature selection was performed using the importance-based RFE method, and the importance was evaluated using the GBM, RF, SVM, and XGB classifiers. When a classifier model is trained on a training dataset, feature weights that reflect the importance of each feature are obtained. After all the features were ranked according to their weights, the feature with the lowest weight value was removed. The classifier is then retrained with the remaining features until there are no more features to learn. Finally, the model-based RFE method can obtain important features and show good performance [20]. The classification models were constructed by RF classifiers using the selected feature values above. As a result of feature reduction using GBM-RFE, RF-RFE, SVM-RFE, and XGB-RFE, the values of accuracy, precision, recall, and F-score for all models were higher than those obtained without RFE. The GBM-RFE showed the highest accuracy and F-score, the GBM-RFE and XGB-RFE had the highest precision, and the RF-RFE had the highest recall. From the results of the receiver operating characteristic curve and the area under the curve (AUC), GBM-RFE had the highest AUC value (Figure 4).

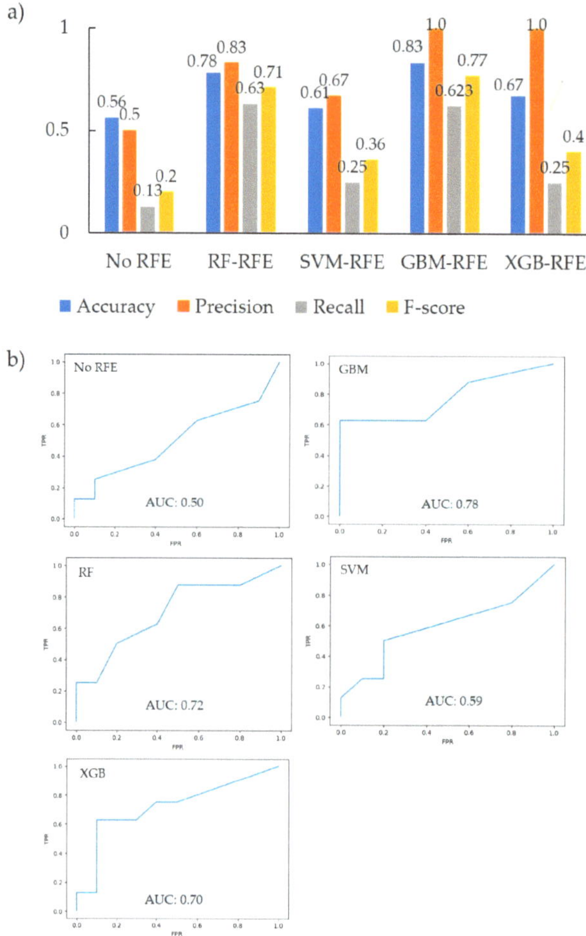

Figure 4. Performance of each model. (**a**) Accuracy, Precision, Recall, and F-score of each RFE; (**b**) ROC and AUC score of each RFE.

The important factors extracted differed for each model (Figure 5). As a feature of RDKit, fr_unbrch_alkane was the top selected feature in all RFE models. For the TD-NMR sequence, several time points of the CPMG, double quantum (DQ) filter, MSE, MAPE, and SE sequences were selected, but most of them were after the intermediate region where the slope of the transition curve becomes gentle (Figure 6). The important factors extracted differed for each model (Figures 5 and 6).

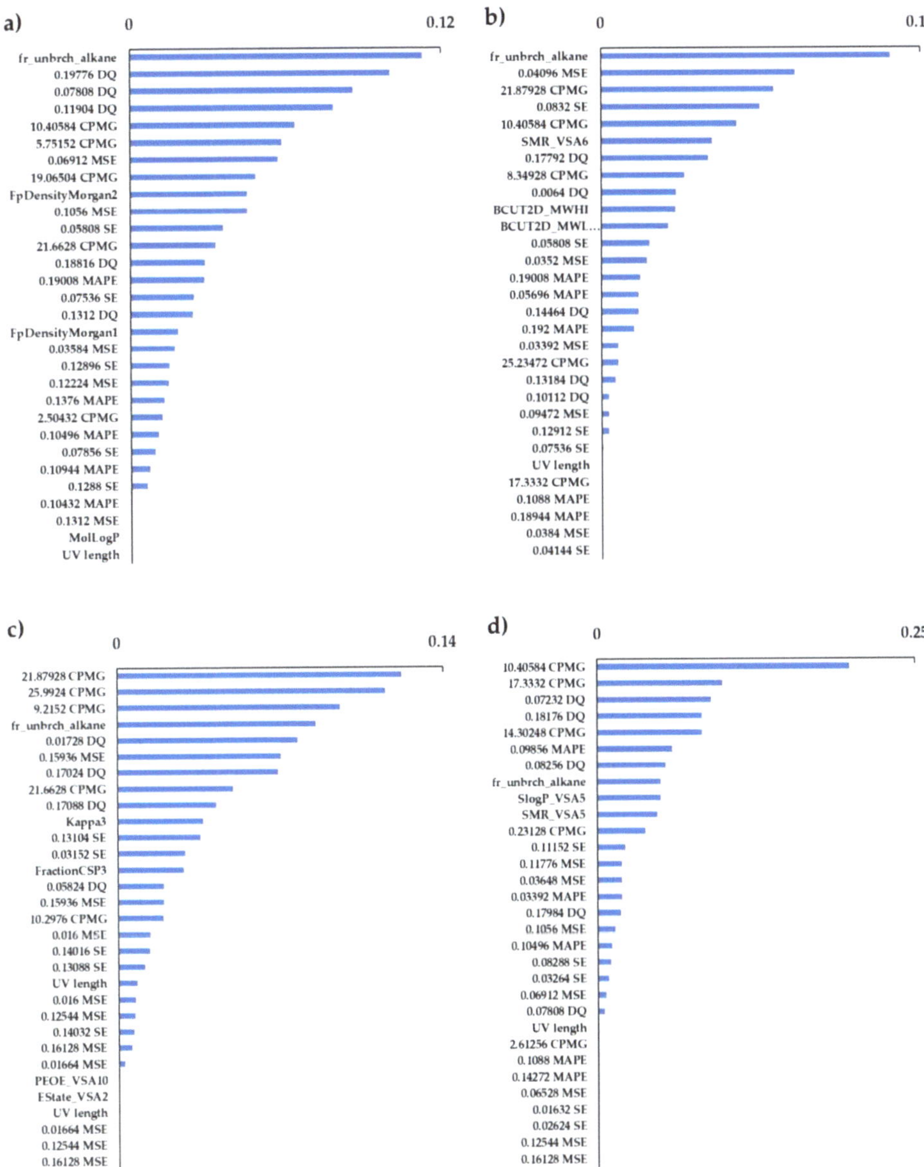

Figure 5. Important features of each RF classification model. Feature selections were performed by (**a**) GBM-RFE, (**b**) RF-RFE, (**c**) SVM-RFE, and (**d**) XGB-RFE.

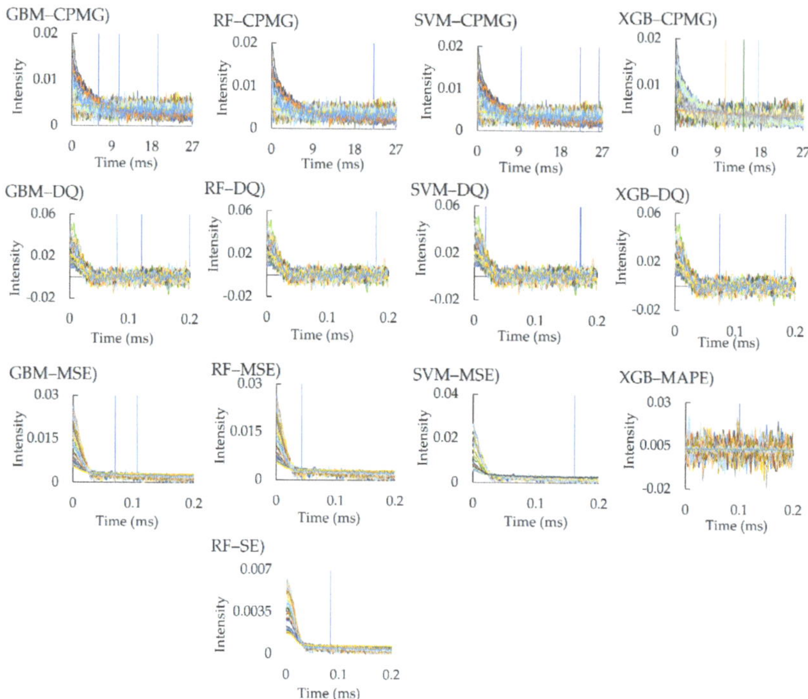

Figure 6. Important features of relaxation time in each pulse sequence. The extracted important features (blue line) for each relaxation time are shown.

4. Discussion

Among the four RFE methods, GBM-RFE exhibited the best feature-selection performance. For LightGBM, following the decision tree analysis, gradient boosting was employed to enhance the accuracy. This boosting technique improves the predictive performance by learning from errors between predicted and actual values; it particularly focuses on data that could not be initially accurately predicted, and this method of growing according to the leaves of a decision tree is called leaf-wise [43].

Although XGB is based on gradient boosting, a difference with the branches of the decision tree, called level-wise, exists for each layer [43]. SVM-RFE, a wrapper method [44], was employed for feature selection in this study using a linear form [45]. However, the linear approach might not have effectively classified the current dataset, which comprised RDKit and five pulse sequence data with diverse characteristics. RF, a bagging method based on decision trees [46], demonstrated the second-best performance among the models. Its strength lies in amalgamating multiple decision trees into an ensemble, which potentially contributes to its effective analysis ability. Although both GBM and XGB are gradient boosting methods, their inherent approaches are different. GBM utilizes the leaf-wise method, focusing on improving accuracy while learning the errors for each leaf. Interestingly, GBM-RFE proved suitable for our dataset with its varying characteristics. Notably, the selection and importance of the features were dependent on the RFE methods employed in the analysis. This dependency underscores the significance of the chosen methodology in determining feature relevance and importance.

TD-NMR measurements offer a wide dynamic range, spanning from sub-microseconds (μs) to seconds, allowing for the extraction of various types of information across different time scales [46]. Mobility within meso-regions, like domain fluctuations, was observed in the μs range [40,47,48], while fluctuations in chain ends or the mobility of unfrozen and bound water were detected at the ms level [31]. The CPMG pulse sequence enables the measurement of long T_2 relaxation times, allowing the analysis of the long components (liquid-like) of polymers [28,49–51]. DQ, MSE, and SE have short T_2 relaxation times and are thus suitable for analysis of the rigid components (solid-like) of polymers [52–54]. The data after the middle region of each transitional curve includes data related to highly motile components. The monomers used in this study contained acrylamide groups, and the monomer FOM-03010 features a betaine structure, which is a zwitterionic group. The C=O and N=H of the acrylamide group [55] (betaine structures [56]) interact with the water surrounding the polymer, forming a hydration layer that inhibits foulant adhesion. In our case, multiple pulse sequences of CPMG, DQ, MSE, and SE were used to detect data from relaxation curves; in addition, various motilities might be involved in the measured contact angle and hydration properties.

Among the several molecular descriptors of physicochemical features, fr_unbrch_alkane (number of unbranched alkanes of at least four members, excluding halogenated alkanes) [57], which represented the proportion of unbranched alkanes, was selected in most of the RFE methods. The presence of branched alkanes in molecular structures is crucial for predicting crystallization owing to their disruptive effect on molecular packing, which destabilizes the liquid crystal phase [58]. This characteristic suggests that hydrophilic monomers may form structures conducive to expressing hydrophilicity by orderly bonding among themselves. Furthermore, molecular fluctuations occur more easily in a linear structure than in a branched structure. Based on the TD-NMR data, intermediate or late relaxation time and high molecular mobility were selected as important factors in the region; thus, we can infer that this molecular mobility (molecular fluctuation) is involved in the expression of hydrophilicity.

As discussed earlier, our current methodology offers the unique advantage of simultaneously providing the essential components from both molecular descriptors and TD-NMR T_2 relaxation curves. These components respectively represent physicochemical and dynamic properties. While understanding both properties is crucial for designing superior surface modifications, the challenge lies in the human capacity to manage a vast array of molecular descriptors and numerous NMR curves obtained through various pulse sequences. An additional noteworthy aspect is our utilization of diverse types of RFE methods. The importance attributed to ML algorithms varies, and relying on a single set of criteria can lead to misunderstandings due to factors like noise or pseudo-correlation. Therefore, our approach is particularly well-suited for a comprehensive and multifaceted examination of material data. In recent years, simple and inexpensive methods using smartphones and ML have been developed to measure contact angles [59,60]. Until recently, expensive contact angle meters were extensively used. However, these new ML-based methods enable the evaluation of the process of creating hydrophilic/hydrophobic polymer coating materials and aid in efficiently managing their manufacturing process. Therefore, incorporating the proposed ML-based method into future studies will be useful.

5. Conclusions

In our study, we showcased feature-selection techniques for molecular descriptors assessed via RDKit and relaxation curves obtained from TD-NMR, employing RFE for surface modifications. Our surface modifications involved copolymerizing various combinations of acrylamide monomers on PET films. To evaluate polymer chain dynamics across a broad time range, we utilized TD-NMR measurements with multiple pulse sequences, standardizing the data obtained from these five sequences. Applying GBM-RFE, RF-RFE, SCM-RFE, and XGB-RFE treatments to these descriptors significantly improved the predictability of the RF classifications. Moreover, our findings highlighted the crucial roles played by both the physicochemical properties and dynamics of polymer chains in determining the surface properties. The RFE method not only enhanced the predictability but also allowed us to extract critical factors or time regions from both physicochemical and TD-NMR data. This deeper insight into underlying mechanisms underscores the versatility of these approaches, extending their applicability beyond surface modification to other materials requiring comprehensive multi-modal information.

Supplementary Materials: The following supporting information can be downloaded at: https:// www.mdpi.com/article/10.3390/polym16060824/s1; Figure S1: chemical structure of monomers; Figure S2: The flow of data analysis of this study; Figure S3: Autoscaling of each T_2 relaxation curve; Figure S4: Performance of each criterion for binary classification; Table S1: list of samples; Table S2: list of the molecular descriptors; Table S3: list of the RDKit descriptors of each monomer; Table S4: the parameter of each random forest classifier model.

Author Contributions: Conceptualization, J.K. and Y.A.; methodology, M.O.; formal analysis, M.O.; writing—original draft, M.O.; writing—review & editing, J.K. and Y.A.; project administration, J.K. All authors have read and agreed to the published version of the manuscript.

Funding: This research received no external funding.

Data Availability Statement: The data underlying this article will be shared on reasonable request to the corresponding author.

Acknowledgments: The authors thank the members of RIKEN for their support with TD-NMR data acquisition and handling.

Conflicts of Interest: The authors declare no conflicts of interest that influence the research reported in this paper.

References

1. Vogelpohl, T.; Töller, A.E. Perspectives on the bioeconomy as an emerging policy field. *J. Environ. Policy Plan.* **2021**, *23*, 143–151. [CrossRef]

2. Reike, D.; Vermeulen, W.J.V.; Witjes, S. The circular economy: New or Refurbished as CE 3.0?—Exploring Controversies in the Conceptualization of the Circular Economy through a Focus on History and Resource Value Retention Options. *Resour. Conserv. Recycl.* **2018**, *135*, 246–264. [CrossRef]

3. Kakadellis, S.; Rosetto, G. Achieving a circular bioeconomy for plastics. *Science* **2021**, *373*, 49–50. [CrossRef] [PubMed]

4. Nasrabadi, A.E.; Ramavandi, B.; Bonyadi, Z. Recent progress in biodegradation of microplastics by *Aspergillus* sp. in aquatic environments. *Colloid. Interface Sci. Commun.* **2023**, *57*, 100754. [CrossRef]

5. Hossain, S.; Manan, H.; Shukri, Z.N.A.; Othman, R.; Kamaruzzan, A.S.; Rahim, A.I.A.; Khatoon, H.; Minhaz, T.M.; Islam, Z.; Kasan, N.A. Microplastics biodegradation by biofloc-producing bacteria: An inventive biofloc technology approach. *Microbiol. Res.* **2023**, *266*, 127239. [CrossRef]

6. Yoshida, S.; Hiraga, K.; Takehana, T.; Taniguchi, I.; Yamaji, H.; Maeda, Y.; Toyohara, K.; Miyamoto, K.; Kimura, Y.; Oda, K. A bacterium that degrades and assimilates poly (ethylene terephthalate). *Science* **2016**, *351*, 1196–1199. [CrossRef]

7. Vuong, P.; McKinley, A.; Kaur, P. Understanding biofouling and contaminant accretion on submerged marine structures. *NPJ Mater. Degrad.* **2023**, *7*, 50. [CrossRef]

8. Qiu, H.; Feng, K.; Gapeeva, A.; Meurisch, K.; Kaps, S.; Li, X.; Yu, L.; Mishra, Y.K.; Adelung, R.; Baum, M. Functional polymer materials for modern marine biofouling control. *Prog. Polymr Sci.* **2022**, *127*, 101516. [CrossRef]

9. Kadoma, Y.; Nakabayashi, N.; Masuhara, E.; Yamauchi, J. Synthesis and Hemolysis Test of the Polymer Containing Phosphorylcholine Groups. *Koubunshi Ronbunshu* **1978**, *35*, 423–427. [CrossRef]

10. Ishihara, K.; Nomura, H.; Mihara, T.; Kurita, K.; Iwasaki, Y.; Nakabayashi, N. Why do phospholipid polymers reduce protein adsorption? *J. Biomed. Mater. Res.* **1998**, *39*, 323–330. [CrossRef]

11. Tanaka, M.; Motomura, T.; Kawada, M.; Anzai, T.; Kasori, Y.; Shiroya, T.; Shimura, K.; Onishi, M.; Mochizuki, A. Blood compatible aspects of poly (2-methoxyethylacrylate) (PMEA)—Relationship between protein adsorption and platelet adhesion on PMEA surface. *Biomaterials* **2000**, *21*, 1471–1481. [CrossRef] [PubMed]

12. Sato, K.; Kobayashi, S.; Kusakari, M.; Watahiki, S.; Oikawa, M.; Hoshiba, T.; Tanaka, M. The Relationship Between Water Structure and Blood Compatibility in Poly (2-methoxyethyl Acrylate) (PMEA) Analogues. *Macromol. Biosci.* **2015**, *15*, 1296–1303. [CrossRef] [PubMed]

13. Ashok, D.; Cheeseman, S.; Wang, Y.; Funnell, B.; Leung, S.F.; Tricoli, A.; Nisbet, D. Superhydrophobic surfaces to combat bacterial surface colonization. *Adv. Mater. Interfaces* **2023**, *10*, 2300324. [CrossRef]

14. Hu, P.; Xie, Q.; Ma, C.; Zhang, G. Silicone-based fouling-release coatings for marine antifouling. *Langmuir* **2020**, *36*, 2170–2183. [CrossRef] [PubMed]

15. Webber, M.J.; Tibbitt, M.W. Dynamic and reconfigurable materials from reversible network interactions. *Nat. Rev. Mater.* **2022**, *7*, 541–556. [CrossRef]

16. Carré, A.; Gastel, J.-C.; Shanahan, M.E.R. Viscoelastic effects in the spreading of liquids. *Nature* **1996**, *379*, 432–434. [CrossRef]

17. Matlahov, I.; van der Wel, P.C. Hidden motions and motion-induced invisibility: Dynamics-based spectral editing in solid-state NMR. *Methods* **2018**, *148*, 123–135. [CrossRef] [PubMed]

18. Gupta, S.; Puttaiahgowda, Y.M.; Parambil, A.M.; Kulal, A. Fabrication of crosslinked piperazine polymer coating: Synthesis, characterization and its activity towards microorganisms. *J. Mol. Struct.* **2023**, *1274*, 134522. [CrossRef]

19. Tsuji, T.; Ono, T.; Taguchi, H.; Leong, K.H.; Hayashi, Y.; Kumada, S.; Okada, K.; Onuki, Y. Continuous Monitoring of the Hydration Behavior of Hydrophilic Matrix Tablets Using Time-Domain NMR. *Chem. Pharm. Bull.* **2023**, *71*, 576–583. [CrossRef]

20. Colnago, L.A.; Wiesman, Z.; Pages, G.; Musse, M.; Monaretto, T.; Windt, C.W.; Rondeau-Mouro, C. Low field, time domain NMR in the agriculture and agrifood sectors: An overview of applications in plants, foods and biofuels. *J. Magn. Reson.* **2021**, *323*, 106899. [CrossRef]

21. Matsumura, T.; Nagamura, N.; Akaho, S.; Nagata, K.; Ando, Y. Spectrum adapted expectation-maximization algorithm for high-throughput peak shift analysis. *Sci. Technol. Adv. Mater.* **2019**, *20*, 733–745. [CrossRef]

22. Bolón-Canedo, V.; Sánchez-Maroño, N.; Alonso-Betanzos, A. Feature selection for high-dimensional data. *Prog. Artif. Intell.* **2016**, *5*, 65–75. [CrossRef]

23. Kusaka, Y.; Hasegawa, T.; Kaji, H. Noise reduction in solid-state NMR spectra using principal component analysis. *J. Phys. Chem. A* **2019**, *123*, 10333–10338. [CrossRef]

24. Peng, W.K.; Ng, T.-T.; Loh, T.P. Machine learning assistive rapid, label-free molecular phenotyping of blood with two-dimensional NMR correlational spectroscopy. *Commun. Biol.* **2020**, *3*, 535. [CrossRef]

25. Florentino-Ramos, E.; Villa-Ruano, N.; Hidalgo-Martínez, D.; Ramírez-Meraz, M.; Méndez-Aguilar, R.; Velásquez-Valle, R.; Zepeda-Vallejo, L.G.; Pérez-Hernández, N.; Becerra-Martínez, E. ^{1}H NMR-based fingerprinting of eleven Mexican Capsicum annuum cultivars. *Food Res. Int.* **2019**, *121*, 12–19. [CrossRef] [PubMed]

26. Zhou, T.; Song, Z.; Sundmacher, K. Big data creates new opportunities for materials research: A review on methods and applications of machine learning for materials design. *Engineering* **2019**, *5*, 1017–1026. [CrossRef]

27. Olfatbakhsh, T.; Andrews, J.L.; Milani, A.S. Materials informatics of woven fabric composites: Effect of different dimensionality reduction and learning methods. *Mater. Today Commun.* **2022**, *32*, 103971. [CrossRef]

28. Yamada, S.; Tsuboi, Y.; Yokoyama, D.; Kikuchi, J. Polymer composition optimization approach based on feature extraction of bound and free water using time-domain nuclear magnetic resonance. *J. Magn. Reson.* **2023**, *351*, 107438. [CrossRef]

29. Forshed, J.; Schuppe-Koistinen, I.; Jacobsson, S.P. Peak alignment of NMR signals by means of a genetic algorithm. *Anal. Chim. Acta* **2003**, *487*, 189–199. [CrossRef]

30. Cho, H.-W.; Kim, S.B.; Jeong, M.K.; Park, Y.; Ziegler, T.R.; Jones, D.P. Genetic algorithm-based feature selection in high-resolution NMR spectra. *Expert. Syst. Appl.* **2008**, *35*, 967–975. [CrossRef] [PubMed]

31. Wei, F.; Fukuchi, M.; Ito, K.; Sakata, K.; Asakura, T.; Date, Y.; Kikuchi, J. Large-scale evaluation of major soluble macromolecular components of fish muscle from a conventional ^{1}H-NMR spectral database. *Molecules* **2020**, *25*, 1966. [CrossRef]

32. Munir, N.; McMorrow, R.; Mulrennan, K.; Whitaker, D.; McLoone, S.; Kellomäki, M.; Talvitie, E.; Lyyra, I.; McAfee, M. Interpretable Machine Learning Methods for Monitoring Polymer Degradation in Extrusion of Polylactic Acid. *Polymers* **2023**, *15*, 3566. [CrossRef] [PubMed]

33. Fallah Atanaki, F.; Behrouzi, S.; Ariaeenejad, S.; Boroomand, A.; Kavousi, K. BIPEP: Sequence-based prediction of biofilm inhibitory peptides using a combination of nmr and physicochemical descriptors. *ACS Omega* **2020**, *5*, 7290–7297. [CrossRef]

34. Jeon, H.; Oh, S. Hybrid-recursive feature elimination for efficient feature selection. *Appl. Sci.* **2020**, *10*, 3211. [CrossRef]

35. Zhang, H.; Zhu, S.; Yang, J.; Ma, A. Advancing strategies of biofouling control in water-treated polymeric membranes. *Polymers* **2022**, *14*, 1167. [CrossRef]

36. Morita, S.; Tanaka, M.; Ozaki, Y. Time-resolved in situ ATR-IR observations of the process of sorption of water into a poly (2-methoxyethyl acrylate) film. *Langmuir* **2007**, *23*, 3750–3761. [CrossRef]

37. Noto, N.; Yada, A.; Yanai, T.; Saito, S. Machine-Learning Classification for the Prediction of Catalytic Activity of Organic Photosensitizers in the Nickel (II)-Salt-Induced Synthesis of Phenols. *Angew. Chem. Int. Ed.* **2023**, *62*, e202219107. [CrossRef]

38. Feng, Z.; Cheng, Y.; Khlyustova, A.; Wani, A.; Franklin, T.; Varner, J.D.; Hook, A.L.; Yang, R. Virtual High-Throughput Screening of Vapor-Deposited Amphiphilic Polymers for Inhibiting Biofilm Formation. *Adv. Mater. Technol.* **2023**, *8*, 2201533. [CrossRef]

39. Ono, S.; Hewage, H.T.; Visvanathan, C. Towards Plastic Circularity: Current Practices in Plastic Waste Management in Japan and Sri Lanka. *Sustainability* **2023**, *15*, 7550. [CrossRef]

40. Schäler, K.; Roos, M.; Micke, P.; Golitsyn, Y.; Seidlitz, A.; Thurn-Albrecht, T.; Schneider, H.; Hempel, G.; Saalwächter, K. Basic principles of static proton low-resolution spin diffusion NMR in nanophase-separated materials with mobility contrast. *Solid. State Nucl. Magn. Reson.* **2015**, *72*, 50–63. [CrossRef]

41. Takamura, A.; Tsukamoto, K.; Sakata, K.; Kikuchi, J. Integrative measurement analysis via machine learning descriptor selection for investigating physical properties of biopolymers in hairs. *Sci. Rep.* **2021**, *11*, 24359. [CrossRef] [PubMed]

42. Weininger, D. SMILES, a chemical language and information system. 1. Introduction to methodology and encoding rules. *J. Chem. Inf. Comput. Sci.* **1988**, *28*, 31–36. [CrossRef]

43. Suenaga, D.; Takase, Y.; Abe, T.; Orita, G.; Ando, S. Prediction accuracy of Random Forest, XGBoost, LightGBM, and artificial neural network for shear resistance of post-installed anchors. *Structures* **2023**, *50*, 1252–1263. [CrossRef]

44. Duan, K.-B.; Rajapakse, J.C.; Wang, H.; Azuaje, F. Multiple SVM-RFE for gene selection in cancer classification with expression data. *IEEE Trans. Nanobiosci.* **2005**, *4*, 228–234. [CrossRef]

45. Ding, Y.; Wilkins, D. Improving the performance of SVM-RFE to select genes in microarray data. *BMC Bioinform.* **2006**, *7*, S12. [CrossRef]
46. Altman, N.; Krzywinski, M. Ensemble methods: Bagging and random forests. *Nat. Meth.* **2017**, *14*, 933–935. [CrossRef]
47. Schlagnitweit, J.; Tang, M.; Baias, M.; Richardson, S.; Schantz, S.; Emsley, L. A solid-state NMR method to determine domain sizes in multi-component polymer formulations. *J. Magn. Reson.* **2015**, *261*, 43–48. [CrossRef] [PubMed]
48. Hara, K.; Yamada, S.; Kurotani, A.; Chikayama, E.; Kikuchi, J. Materials informatics approach using domain modelling for exploring structure-property relationships of polymers. *Sci. Rep.* **2022**, *12*, 10558. [CrossRef] [PubMed]
49. Borgia, G.; Fantazzini, P.; Ferrando, A.; Maddinelli, G. Characterisation of crosslinked elastomeric materials by ^{1}H NMR relaxation time distributions. *Magn. Reson. Imag.* **2001**, *19*, 405–409. [CrossRef]
50. Dare, D.; Chadwick, D. A low resolution pulsed nuclear magnetic resonance study of epoxy resin during cure. *Int. J. Adhes. Adhes.* **1996**, *16*, 155–163. [CrossRef]
51. Li, J.; Ma, E. Characterization of water in wood by time-domain nuclear magnetic resonance spectroscopy (TD-NMR): A review. *Forests* **2021**, *12*, 886. [CrossRef]
52. Grunin, L.; Ivanova, M.; Schiraya, V.; Grunina, T. Time-Domain NMR Techniques in Cellulose Structure Analysis. *Appl. Magn. Reson.* **2023**, *54*, 929–955. [CrossRef]
53. Garcia, R.H.d.S.; Filgueiras, J.G.; Colnago, L.A.; de Azevedo, E.R. Real-Time Monitoring Polymerization Reactions Using Dipolar Echoes in 1H Time Domain NMR at a Low Magnetic Field. *Molecules* **2022**, *27*, 566. [CrossRef]
54. Uguz, S.S.; Ozel, B.; Grunin, L.; Ozvural, E.B.; Oztop, M.H. Non-Conventional Time Domain (TD)-NMR Approaches for Food Quality: Case of Gelatin-Based Candies as a Model Food. *Molecules* **2022**, *27*, 6745. [CrossRef] [PubMed]
55. Futscher, M.H.; Philipp, M.; Müller-Buschbaum, P.; Schulte, A. The Role of Backbone Hydration of Poly(N-isopropyl acrylamide) Across the Volume Phase Transition Compared to its Monomer. *Sci. Rep.* **2017**, *7*, 17012. [CrossRef]
56. Ishihara, K.; Mu, M.; Konno, T.; Inoue, Y.; Fukazawa, K. The unique hydration state of poly(2-methacryloyloxyethyl phosphorylcholine). *J. Biomater. Sci. Polym. Ed.* **2017**, *28*, 884–899. [CrossRef]
57. Shamsara, J. A random forest model to predict the activity of a large set of soluble epoxide hydrolase inhibitors solely based on a set of simple fragmental descriptors. *Comb. Chem. High. Throughput Screen.* **2019**, *22*, 555–569. [CrossRef] [PubMed]
58. Chen, C.-H.; Tanaka, K.; Kotera, M.; Funatsu, K. Comparison and improvement of the predictability and interpretability with ensemble learning models in QSPR applications. *J. Cheminf.* **2020**, *12*, 1–16. [CrossRef]
59. Kabir, H.; Garg, N. Machine learning enabled orthogonal camera goniometry for accurate and robust contact angle measurements. *Sci. Rep.* **2023**, *13*, 1497. [CrossRef]
60. Chen, H.; Muros-Cobos, J.L.; Amirfazli, A. Contact angle measurement with a smartphone. *Rev. Sci. Instrum.* **2018**, *89*, 035117. [CrossRef]

 *polymers*

Article

Spatial-Temporal Kinetic Behaviors of Micron-Nano Dust Adsorption along Epoxy Resin Insulator Surfaces and the Physical Mechanism of Induced Surface Flashover

Naifan Xue [1,*], Bei Li [2], Yuan Wang [1], Ning Yang [1], Ruicheng Yang [1], Feichen Zhang [1] and Qingmin Li [1]

[1] State Key Laboratory of Alternate Electrical Power System with Renewable Energy Sources, North China Electric Power University, Beijing 102206, China; wy_vera0209@163.com (Y.W.); 16608030457@163.com (N.Y.); yrc20000810@163.com (R.Y.); zfc020306@outlook.com (F.Z.); lqmeee@ncepu.edu.cn (Q.L.)
[2] School of Electrical Engineering, Hebei University of Technology, Tianjin 300130, China; 2022moltres@gmail.com
* Correspondence: 18612332728@163.com; Tel.: +86-18612332728

Abstract: The advanced Gas Insulated Switchgear/Gas Insulated Lines (GIS/GIL) transmission equipment serves as an essential physical infrastructure for establishing a new energy power system. An analysis spanning nearly a decade on faults arising from extra/ultra-high voltage discharges reveals that over 60% of such faults are attributed to the discharge of metal particles and dust. While existing technical means, such as ultra-high frequency and ultrasonic sensing, exhibit effectiveness in online monitoring of particles larger than sub-millimeter dimensions, the inherent randomness and elusive nature of micron-nano dust pose challenges for effective characterization through current technology. This elusive micron-nano dust, likely concealed as a latent threat, necessitates special attention due to its potential as a "safety killer". To address the challenges associated with detecting micron-nano dust and comprehending its intricate mechanisms, this paper introduces a micron-nano dust adsorption experimental platform tailored for observation and practical application in GIS/GIL operations. The findings highlight that micron-nano dust's adsorption state in the electric field predominantly involves agglomerative adsorption along the insulator surface and diffusive adsorption along the direction of the ground electrode. The pivotal factors influencing dust movement include the micron-nano dust's initial position, mass, material composition, and applied voltage. Further elucidation emphasizes the potential of micron-nano dust as a concealed safety hazard. The study reveals specific physical phenomena during the adsorption process. Agglomerative adsorption results in micron-nano dust speckles forming on the epoxy resin insulator's surface. With increasing voltage, these speckles undergo an "explosion", forming an annular dust halo with deepening contours. This phenomenon, distinct from the initial adsorption, is considered a contributing factor to flashovers along the insulator's surface. The physical mechanism behind flashovers triggered by micron-nano dust is uncovered, highlighting the formation of a localized short circuit area and intense electric field distortion constituted by dust speckles. These findings establish a theoretical foundation and technical support for enhancing the safe operational performance of AC and DC transmission pipelines' insulation.

Keywords: epoxy resin insulator; micron-nano dust; GIS/GIL; adsorption behavior; surface flashover

Citation: Xue, N.; Li, B.; Wang, Y.; Yang, N.; Yang, R.; Zhang, F.; Li, Q. Spatial-Temporal Kinetic Behaviors of Micron-Nano Dust Adsorption along Epoxy Resin Insulator Surfaces and the Physical Mechanism of Induced Surface Flashover. *Polymers* **2024**, *16*, 485. https://doi.org/10.3390/polym16040485

Academic Editor: Ilker S. Bayer

Received: 5 January 2024
Revised: 28 January 2024
Accepted: 6 February 2024
Published: 9 February 2024

1. Introduction

The establishment of a new power system integrating multimodal energy sources is an important technical way to realize carbon emission targets, while advanced transmission and transformation equipment is an indispensable physical support for the construction of a new power system. Gas Insulated Switchgear/Gas Insulated Lines (GIS/GIL) are playing crucial roles as the "network joints" and "security guards" of energy transmission [1,2]. GIS/GIL, widely utilized in ultra/high voltage power grids, boasts unique advantages such as minimal transmission loss, substantial capacity, high operational reliability, and integrated transmission of renewable energy sources [3–5]. Despite these merits, insulation failure remains the primary cause of GIS/GIL operational issues, with internal metal particle discharge emerging as a key safety concern.

Metal particles are easily moved to the vicinity of the insulators inside the GIS/GIL equipment due to the electric field, making the electric field distortion near the fragile insulator triple junction more serious. Because epoxy resin has good heat resistance and electrical insulation, internally, GIS/GIL mainly use insulators cast from epoxy resin composite materials. Under the electric field, the epoxy resin surface provides an attachment environment for metal particles or micro-nano dust, and the metal contaminants adhering to the epoxy resin surface, or doing the firefly motion near the epoxy resin surface, greatly increase the probability of electrical insulation failure.

For discharge faults triggered by large metal particles (above the submillimeter level), existing UHF and ultrasonic sensing technologies offer acceptable/better online monitoring and early warning capabilities. However, a significant number of unexplained insulation discharge faults persist at engineering sites. Even with current UHF and ultrasonic sensing technologies, better online monitoring and early warning for discharge faults caused by large-size metal particles (sub-millimeter level or larger) are achievable [6–8], yet numerous unexplained insulation discharge faults still occur in engineering settings. Through comprehensive analysis, micron-nano dust emerges as a likely "safety killer" with characteristics of high randomness, insidiousness, and inevitability. The smaller size of micron-nano dust, coupled with its strong physical and chemical activity, pronounced randomness in movement, and clustering effects, facilitates its migration near insulators induced by charge dynamics transfer, electromagnetic radiation, acousto-optic radiation, and distinctive statistical characteristics of larger particles [9–11]. Unfortunately, existing UHF and ultrasonic sensing technologies prove ineffective in detecting the dispersed motion and discharge signals of nano dust, further hindering the characterization of its electrodynamic properties.

Ruixue Liang [12] investigated the adsorption behavior of micron-sized metal dust, identifying three modes of motion near the insulator: accumulation, diffusion, and others. Notably, the experiment took place in ambient air with PTFE insulator material. While some disparity exists with real-world engineering conditions, Liang's work lays a foundational understanding for studying micron-scale dust movement characteristics. Jingrui Wang [13] explored the distribution characteristics of metal dust adsorption near insulators with varying inclination angles, revealing that higher angles correlate with reduced dust accumulation. However, limitations arose from simultaneously measuring dust adsorption and flashover breakdown voltage, introducing substantial error, and lowering experimental reliability. In another study, Zhang Lian-gen [14] delved into partial discharge induced by millimeter-level metal foreign matter on insulator surfaces, describing the persistent erosion caused by dense micro-discharge patterns under constant voltage. Yet, the adsorption and discharge mechanisms of micrometer-level metal foreign matter were left unexplored. Xu Yuan [15] summarized the motion characteristics of different quantities of 100-micron metal particles on GIS insulator surfaces, qualitatively analyzing correlations with partial discharge characteristics but without establishing direct links.

Considering the studies, while it is recognized that fine metal dust in GIS/GIL readily adsorbs onto insulator surfaces, the kinetic behavior and adsorption mechanism remain unclear. Building on existing research on the motion behavior and discharge characteristics of millimeter-sized metal particles, this paper establishes a metal powder adsorption experimental platform suitable for GIS/GIL observation and operation. The focus is on the impact of aberrant electric fields and microscopic force fields on metal dust kinetic behavior patterns. The investigation aims to develop a multi-physical force analysis model for metal dust, unraveling the adsorption mechanism near GIS/GIL basin insulators. This research provides a fundamental basis for dust protection and insulation design.

2. Experimental Platform and Parameter Setting

To elucidate the adsorption dynamics of micronized dust along the gas-solid interface of insulators in GIS/GIL, we devised a schematic diagram depicting the coaxial cylindrical electrode experimental platform, as illustrated in Figure 1. The experimental setup comprises three main components: the experimental chamber housing the coaxial cylindrical model, the DC power source, and the data processing unit. The actual DC power supply and experimental chamber configuration are detailed in Figure 2.

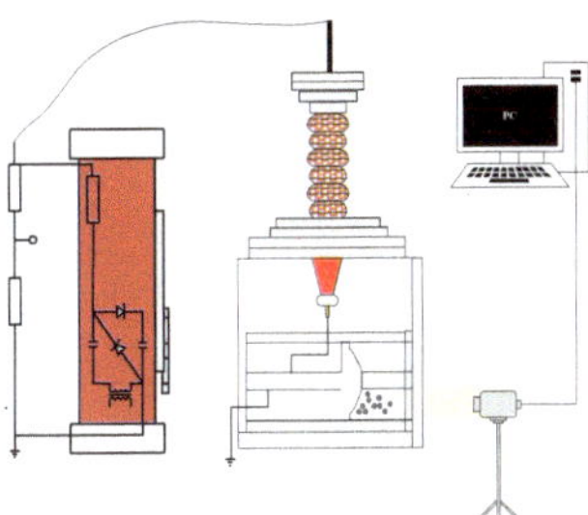

Figure 1. Schematic diagram of the experimental platform for coaxial cylindrical electrodes.

Figure 2. True type power supply and experimental chamber.

The coaxial cylindrical model, depicted in Figure 3, is a scaled representation comprising a high-voltage electrode and a ground electrode. The high-voltage electrode is an aluminum cylinder measuring 40 mm in diameter and 260 mm in length, while the ground electrode is a cylindrical aluminum shell with an inner diameter of 120 mm, an outer diameter of 140 mm, and a length of 260 mm. Both electrodes are supported by a transparent acrylic plate, ensuring their center axes coincide. The coaxial cylindrical electrode platform employs a scaled basin insulator featuring an inner diameter of 40 mm, an outer diameter of 120 mm, a thickness of 40 mm, and an inclination angle of 45 degrees. The chemical composition of the scaled basin insulator used in the experiment includes epoxy resin, alumina, and a curing agent. Bisphenol A diglycidyl ether (DGEBA) was used as the epoxy monomer, and methyl tetrahydrophthalic anhydride (MTHPA) was used as the curing agent. The mass ratio between the components was epoxy resin 80:alumina 100:curing agent 330. The curing condition was as follows: cure at 100 °C for 4 h, then increase the temperature, and then continue to cure at 150 °C for 10 h. Alumina is the main material to improve the performance of epoxy resin because of its high mechanical strength and strong chemical stability. Adding alumina can effectively improve the impact strength, bending strength, thermal conductivity, and glass transition temperature of epoxy insulating materials, and to a certain extent, increase the resistivity and improve the flashover resistance. Moreover, the alumina material could resist the corrosion of sulfur hexafluoride decomposition gas. Figure 4 provides three perspectives of the coaxial cylindrical scaled model.

Figure 3. Scaling model of a coaxial cylindrical electrode.

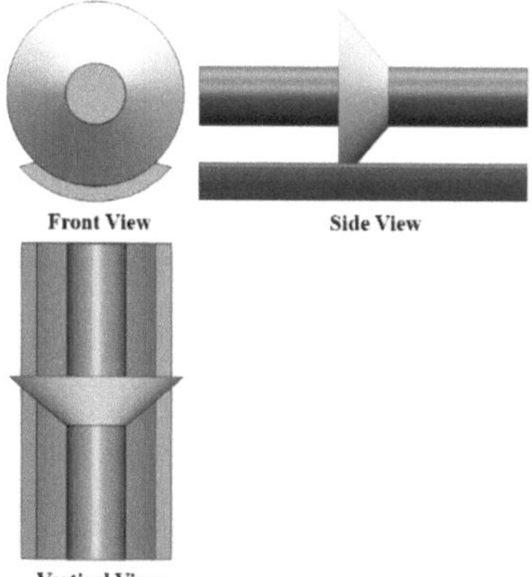

Figure 4. Three views of a coaxial cylindrical electrode.

The high-voltage electrode connects to the DC power supply via a high-voltage wire within the experimental chamber, while the ground electrode links to the ground wire, establishing a comprehensive electrical circuit. Throughout the experiment, the coaxial cylindrical electrode is positioned within a square closed experimental chamber, featuring a transparent observation window on one side. A supplementary light source enhances the overall chamber illumination, and a high-definition camera, positioned on the opposing side, records the dust adsorption process.

To validate the equivalence of the electric field in the semi-enclosed coaxial cylindrical experimental platform and the fully enclosed coaxial cylindrical real platform of GIS/GIL, finite element simulation is employed. Simulation models for semi-enclosed and fully enclosed coaxial columns are developed, with the electric field distribution at the gas-solid interface obtained under a 25 kV voltage in an SF6 gas environment. Given the placement of dust at the bottom of the insulator's convex side during the experiment, the dust adsorption area aligns roughly within 45 degrees of the gas-solid interface corresponding to the ground electrode. The simulation results in Figure 5 indicate that the maximum field strength in Figure 5a is 3.74 kV/mm, and in Figure 5b it is 3.12 kV/mm, with the size of the field distribution being essentially identical. The maximum field strength near the surface of the epoxy resin in a 126 kV true GIS/GIL is around 3.5 kV/mm. This confirms that the coaxial cylindrical electrodes employed in the experiment are equivalent to the SF6 gas simulation platform, affirming the effectiveness and validity of the experimental setup.

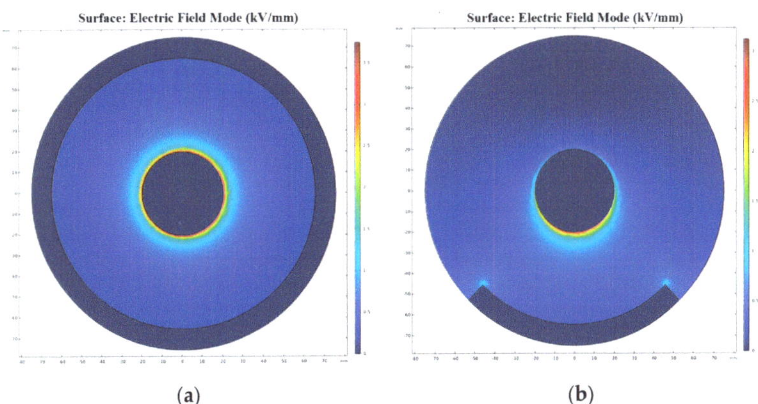

Figure 5. E-field distribution of the test rigs. (**a**) Fully enclosed chamber; (**b**) Semi-closed chamber.

The experimental setup positioned the dust, as depicted in Figure 6, on top of the convex insulator's ground electrode. Figure 6 is the vertical view in Figure 4. The blue dotted line represents the top view profile of the insulator, and the red dotted line represents the top view profile of the high-voltage electrode. The placement was segmented into a nine-grid configuration, with combinations in positions 1–3 and 7–9 representing the furthest distances among the three combinations. Before initiating the experiment, aluminum powder underwent more than 24 h of drying in an oven to ensure complete dryness, mitigating the potential impact of moisture on experimental outcomes. To minimize the influence of prior experiments on subsequent iterations, a thorough wipe-down of the high-voltage electrode, ground electrode, insulator surface, and internal cavity with ethanol preceded each new round of experiments. This step aimed to eliminate surface charge effects, with experiments commencing after the ethanol had evaporated. Precise control was maintained using a high-precision balance (with an error not exceeding 0.1%) to measure the required dust quantity. The experiments were conducted at a temperature of 25 °C, within a 0.1 MPa SF6 environment, with a boosting speed of 2 kV/s. Each set of experiments underwent five repetitions to calculate an average value, thus reducing uncertainties arising from accidental phenomena in the experimental results.

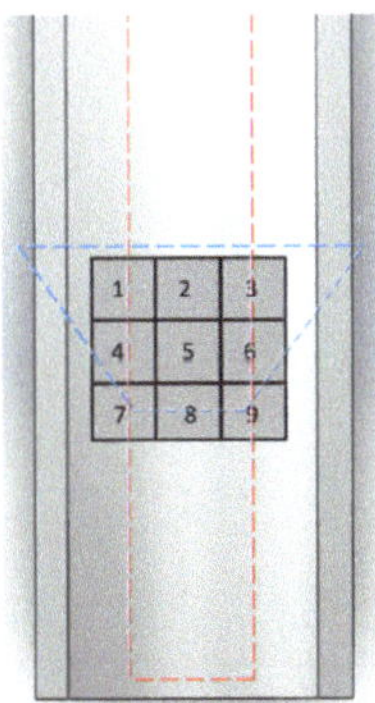

Figure 6. Initial positions of dust.

The typical forces of micron–nano dust are shown in Table 1, including gravity G, buoyancy F_b, electrical force F_E, coulomb force F_q, van der Waals force between two contiguous dusts F_{vdW}, van der Waals force between two untouched dusts F, van der Waals force between dust and surface F_{Rumpf}, and adhesive force F_{adhe}.

Table 1. Typical forces of dust.

Typical Force	Formula	Typical Force	Formula
Gravity	$G = \frac{4}{3}\pi a^3 \rho_{dust} g$		$F_{vdW} = \frac{3}{2}\pi w a$
Buoyancy	$F_b = \frac{4}{3}\pi a^3 \rho_{gas} g$	Van der Waals force	$F = \frac{2\pi w a}{3}\left[\frac{1}{4}\left(\frac{h}{z_0}\right)^{-8} - \left(\frac{h}{z_0}\right)^{-2}\right]$
Electric force	$F_E = qE = q\dfrac{U_{dc}}{r\ln\frac{R_2}{R_1}}$		$F_{Rumpf} = \frac{A}{6}\left[\frac{r_0 a}{z_0^2(a+r_0)} + \frac{a}{(z_0+r_0)^2}\right]$
Coulomb force	$F_q = \frac{1}{4\pi\varepsilon_0}\frac{q_1 q_2}{(2a+h)^2}$	Adhesive force	$F_{adhe} = w\frac{\pi}{180}a\arccos(1 - \frac{4}{3}\pi a^3 \rho_{Al} g k_{tension})$

3. Kinetic Behavior of Micron-Nano Dust Adsorption

3.1. General Adsorption Behavior of Micron-Nano Dust

A uniform deposition of 20 mg of 50 nm aluminum dust occurs on the ground electrode, aligned parallel to the high-voltage conductor at position 258. A negative polarity DC voltage is applied, following the procedures outlined in the initial section, to observe the real-time adsorption movement behavior of micron-nano dust along the insulator surface as voltage increases. The position of the insulator housing the dust is captured in Figure 7. Notably, the 50 nm aluminum dust appears silver-black due to optical dispersion effects. Experimental findings indicate that micron-nano dust gradually begins to lift from -10 kV. In contrast to larger particles, the lifting of micron-nano dust is a continuous process, with not all particles lifting simultaneously. Therefore, this paper focuses on recording the starting lifting voltage of micron-nano dust. As the voltage increases progressively, the movement of micron-nano dust in the electric field primarily manifests in two forms: agglomerative adsorption along the insulator interface and diffusive adsorption along the direction of the ground electrode, as depicted in Figure 8.

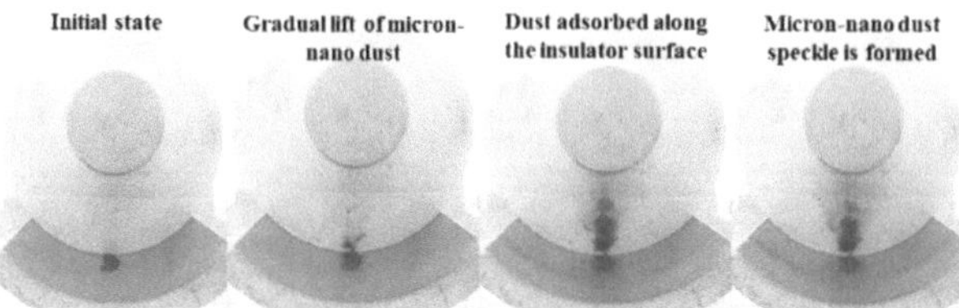

Figure 7. Adsorption behavior of micron-nano aluminum dust parallel to a high-voltage conductor (position 258).

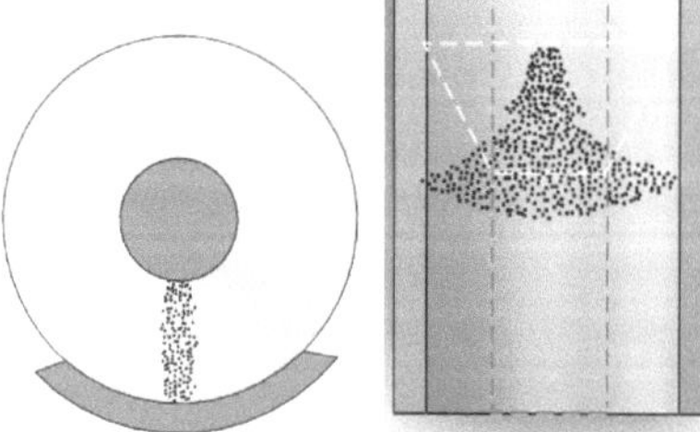

Figure 8. Agglomerative adsorption along the insulator surface, diffusive adsorption in the direction of the ground electrode.

Upon reaching voltages beyond -10 kV, micron-nano dust undergoes elevation and migrates towards the insulator surface. With a continuous increase in voltage, the adsorption quantity of micron-nano dust progressively rises, leading to darker and more substantial agglomerations of micron-nano dust. At -40 kV, agglomerated micron-nano dust forms distinctive patches and lumps on the convex insulator surface, herein referred to as micron-nano dust speckles. These speckles exhibit stable adhesion to the insulator surface, with the adsorption state and quantity remaining relatively constant if the voltage increase is halted. However, if voltage escalation persists, existing micron-nano dust speckles at the gas-solid interface emerge as pivotal factors inducing subsequent flashovers along the surface of the epoxy resin insulator.

Regarding the diffusion and adsorption behavior of micron-nano dust along the direction of the ground electrode, when the voltage surpasses -15 kV, both the diffusion and adsorption area of micron-nano dust expand, accompanied by an increase in diffusion and adsorption quantity. A penetrating dust band forms perpendicular to the direction of the high-voltage conductor on the ground electrode's surface. The color of this dust band intensifies with the gradual voltage increase, resembling a "tree-like" structure in a top-view perspective. It is noteworthy that, considering solely the movement of dust adsorption, after the formation of micron-nano dust patches and the completion of diffusive adsorption on the ground electrode, a surplus of the initial 20 mg of micron-nano dust persists. Subsequent experiments during this period do not observe the phenomenon of complete adsorption of the initial dust. This remaining pile of micron-nano dust provides ample metal reactants for subsequent discharge chemical reactions triggered by flashovers along the surface.

3.2. Key Factors Affecting the Behavior of Micron-Nano Dust Motions

Experiments involving the adsorption of 20 mg of 50 nm aluminum dust reveal a direct correlation between the initial state of the dust and its subsequent adsorption behavior along the epoxy resin insulator. This paper systematically identifies and summarizes four key factors influencing the adsorption behavior of micron-nano dust: the initial position of the dust, its initial mass, the type of electric current (AC/DC), and the material composition of the dust.

3.2.1. Adsorption Behavior of Micronized Dust at Different Initial Positions

In the experiment depicted in Figure 7, micron-nano dust is arranged parallel to the high-voltage conductor. At position 258, the micron-nano dust is positioned directly below the high-voltage conductor. Using this position as a reference point, adsorption experiments are conducted with 20 mg of 50 nm aluminum dust along the surface of the epoxy resin insulator. The dust is arranged at position 147, parallel to the high-voltage conductor and below the side, at position 456, perpendicular to the direction of the high-voltage conductor, at position 159, across the arrangement area, and at the dual initial position 147/369, parallel to the high-voltage conductor. The dynamic aspects of the adsorption behavior are illustrated in Figure 9.

The complete adsorption processes of micron-nano dust at the initial positions of 147, 456, 159, and 147/369 closely mirror that of the dust at position 258, as depicted in Figure 9. The 50 nm aluminum dust gradually lifts from -12 kV, -10 kV, -15 kV, and -10 kV, adhering radially upward along the coaxial cylindrical electrodes, exhibiting an upward-rightward slanting adsorption state from the main view. As the voltage progressively increases, the quantity of inclined agglomerated micron-nano dust adsorption rises, accompanied by a deepening color. By the time the voltage reaches -40 kV, -42 kV, -42 kV, and -44 kV, stable tilted micron-nano dust speckles form on the surface of the epoxy insulator. Concerning the diffuse adsorption behavior on the ground electrode, the overall development process mirrors that of the dust at position 258. Taking the experiment based on position 147 as an example, when the voltage surpasses -18 kV, the dust simultaneously disperses along the ground electrode, away from the insulator's direction and in the direction of the vertical high-voltage conductor. Since the initial position is on the left side of the ground electrode, the resulting diffusion traces exhibit distinct left-deep-right-shallow characteristics.

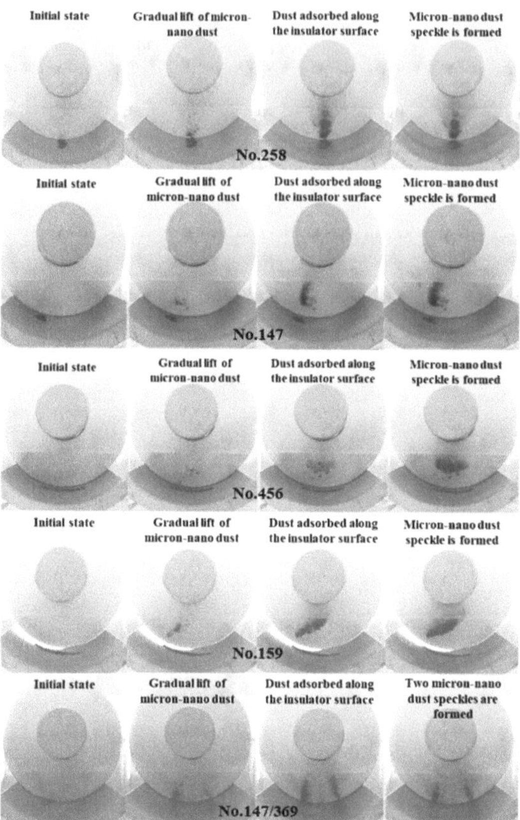

Figure 9. Micron-nano dust adsorption behaviors under different initial circumstances.

Analysis reveals that the adsorption behaviors of the two types of micron-nano dust, agglomerated and diffused along the ground electrode arranged parallel to the high-voltage conductor (positions 258 and 147), are similar. The deviation in the corresponding voltage values of the dynamic process is relatively small. This proximity arises from the modest difference in the required climb height for the lifting of 50 nm aluminum dust at the two initial positions, thus leading to closely aligned voltage values essential for the formation of analogous adsorption characteristics. In contrast to the initial two sets of experiments, the length of the transverse micron-nano dust patches along the radial direction of the insulator is notably shorter. This variation presents the possibility of subsequently impeding the development of flashovers along the insulator surface. Although the deviation in voltage values corresponding to the dynamic adsorption process is minimal, it is closely aligned with that of the formation of dust patches on parallel high-voltage conductors. Due to the extensive distribution of micron-nano dust arranged according to an initial position on the surface of the ground electrode (position 159), the time and voltage required from the start of lifting to the stable micron-nano dust patch formation are higher, by 25%, than that of the equivalent mass of dust starting at position 258.

Figure 9 also illustrates the adsorption process of micron-nano dust at dual initial positions, specifically chosen parallel to the high-voltage conductor at positions 147 and 369. It is noteworthy that the total mass of the two piles of micron-nano dust remains at 20 mg. In comparison with the single micron-nano dust speckle formed by agglomerative adsorption at position 147, the double micron-nano dust speckle formed is noticeably lighter in color and smaller in adsorption area.

While the process of dust speckle formation for different initial positions of micron-nano dust exhibited minimal differences, there was a significant variation in the final morphology of the formed micron-nano dust speckle, as well as the time and voltage required. Therefore, the initial position of micron-nano dust emerges as a crucial factor influencing the morphology of micron-nano dust speckles formed through agglomerative adsorption.

3.2.2. Adsorption Behavior of Micron-Nano Dust with Different Initial Masses

From the above analysis of the double micron-nano dust speckles shown in Figure 9, it is evident that the double micron-nano dust speckle is notably lighter in color and has a smaller adsorption area compared to the single micron-nano dust speckle formed by agglomerative adsorption at location 147. This difference arises because the double initial dust pile is smaller in mass than the single pile, and the total amount utilized for agglomerative adsorption is less than that of the single dust pile.

Figure 10 illustrates the adsorption process of 40 mg of 50 nm micron-nano aluminum dust with the initial position located in the direction of the parallel high-voltage conductor (position 258). The dust gradually lifts from -10 kV, and the agglomerative adsorption process does not significantly differ from the dynamic adsorption process shown in Figure 7. The dust speckle is also formed at -40 kV, but the surface area of the micron-nano dust speckle formed by the 40 mg dust is larger, and the color is darker. Regarding the diffusive adsorption behavior of micron-nano dust along the direction of the ground electrode, the higher the mass, the larger the diffusive adsorption area. Due to the limited size of the ground electrode used in the experiments, most of the dust diffused to the area outside the ground electrode.

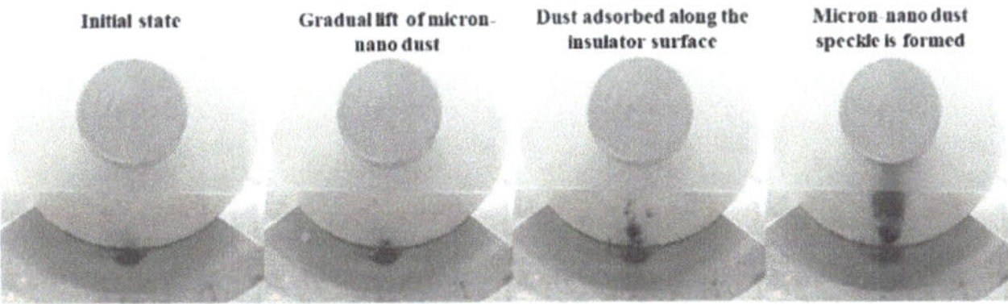

Figure 10. Adsorption behavior of 40 mg of micronized dust at position 258.

Simultaneously, experiments conducted with varying initial masses at the same position reveal a more pronounced agglomerative adsorption phenomenon with an increase in the initial mass. For micron-nano dust positioned at location 258, when its initial mass is 40 mg or greater, the insulator surface experiences an intense along-face flashover phenomenon shortly after the formation of a micron-nano dust speckle, as depicted in Figure 11. Considering that the micron-nano dust speckle induces a flashover breakdown voltage of -40 kV along the surface, equivalent to the air breakdown voltage of the test system, it can be postulated that micron-nano dust with an initial mass greater than or equal to 40 mg, situated in position 258, will lead to a 0.1 MPa SF6 failure.

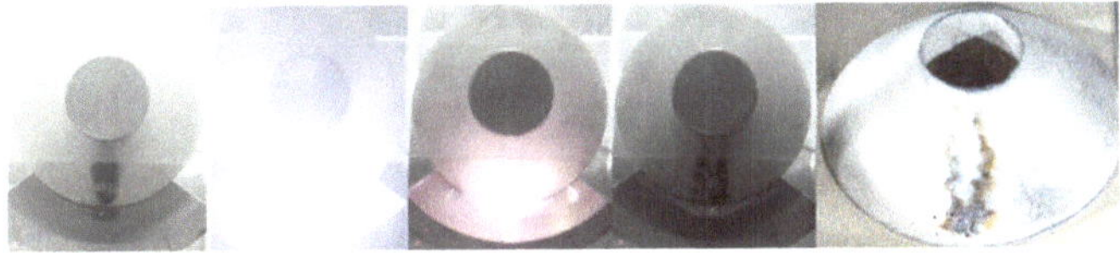

Figure 11. Micron-nano dust speckle induced a flashover along the surface when the initial dust mass was greater than or equal to 40 mg.

Based on the experimental phenomena and analysis, it is apparent that the dynamic process of dust patch formation remains fundamentally consistent across different initial masses of micron-nano dust. However, there is a substantial disparity in the final formation of micron-nano dust patches in terms of surface area, and when the mass surpasses a critical threshold, these dust patches significantly diminish the insulation margin of the gas-solid interface, directly leading to flashovers along the surface. Consequently, the initial mass emerges as a crucial factor influencing the behavior of micron-nano dust movement.

3.2.3. Adsorption Behavior of Micronized Dust When Externally Applied Voltage Is AC

Figure 12 illustrates the adsorption process of a 20 mg, 50 nm micron-nano aluminum speckle under externally applied AC voltage, with the initial position situated in the direction of the parallel high-voltage conductor (position 258). The micron-nano dust initiates at a voltage of 10 kV, and the corresponding voltage for the formation of the micron-nano dust speckle is 42 kV. It is noteworthy that the stability of the micron-nano dust speckle formed under AC voltage is inferior to that under DC voltage. Concurrently, the diffusion and adsorption behavior of the ground electrode differs significantly. High-definition camera observations reveal that the micron-nano dust generates an "air wave" on the surface of the ground electrode, oscillating back and forth with the alternating electric field in the direction of the parallel high-voltage conductor.

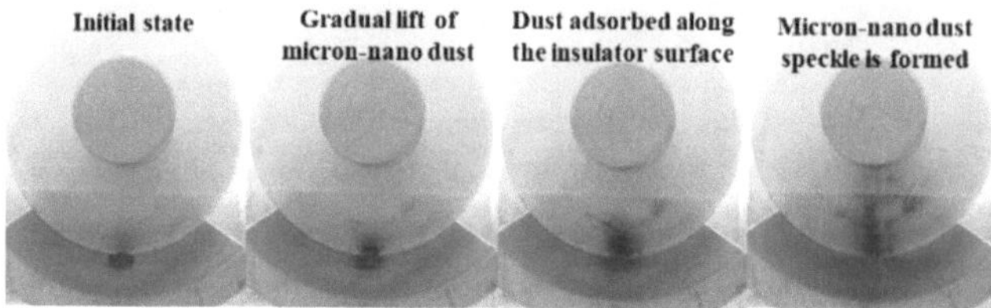

Figure 12. Adsorption behavior 20 mg of micron-nano dust at position 258 under AC voltage.

3.2.4. Adsorption Behavior of Micron-Nano Dust from Different Materials

The adsorption experiments involved the placement of 100 mg of 50 nm copper dust at position 258 in the direction of the parallel high-voltage conductor, as depicted in Figure 13. The behavior of the 50 nm copper dust is observed as it gradually lifts from −14 kV and adsorbs radially upward along the coaxial cylindrical electrode. With the incremental increase in voltage, there is a gradual augmentation in the amount of tilted, agglomerated micron-nano dust adsorption, accompanied by a deepening color. By the time the voltage reaches 40 kV, stable, tilted micron-nano dust spots manifest on the surface of the epoxy insulator. Regarding the diffuse adsorption behavior on the ground electrode, the overall developmental process mirrors that of the dust at position 258. Once the voltage surpasses −10 kV, the dust spreads concurrently along the ground electrode, extending away from the insulator and perpendicular to the high-voltage conductor. This results in a ground electrode diffusion trace that is notably deeper than the trace formed by the aluminum dust.

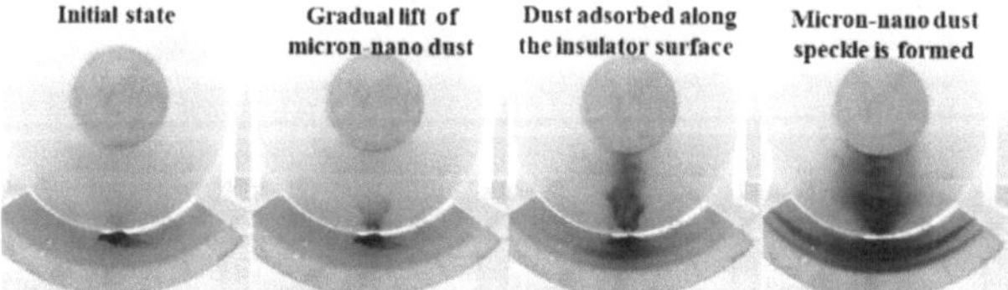

Figure 13. Adsorption behavior of micron-nano copper dust at a parallel high-voltage conductor (position 258).

In summary, as the voltage steadily rises, the adsorption state of micron-nano dust in the electric field primarily involves agglomerative adsorption along the insulator surface and diffusive adsorption along the direction of the ground electrode. The pivotal factors influencing the motion behavior encompass the initial position of the micron-nano dust, its mass, material composition, and the externally applied voltage.

4. Dynamic Processes Prior to Flashover along the Surface Induced by Attached Micron-Nano Dusts

To investigate the physical mechanism of micron-nano dust-induced flashover along the gas-solid interface of epoxy resin insulators, we extended our observations from the micron-nano dust adsorption behavior experiments mentioned above. We systematically increased the voltage to examine the flashover induced by the micron-nano dust speckle formed along the surface.

4.1. Special Physical Phenomena Prior to Flashover along the Surface

As depicted in Figure 7, during the adsorption experiment with 20 mg of 50 nm aluminum dust starting at position 258, stable micron-nano dust speckles formed on the epoxy insulator's surface when the voltage reached −40 kV. As we further increased the voltage, a unique physical phenomenon, distinct from the initial lift adsorption, emerged, as illustrated in Figure 14.

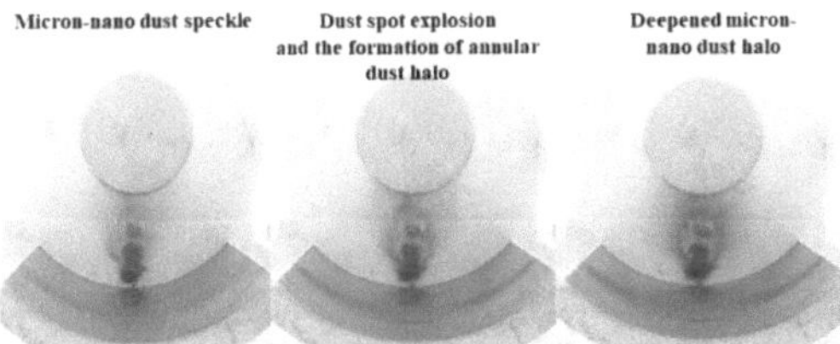

Figure 14. Special physical phenomena of 20 mg 50 nm aluminum dust speckle at position 258.

Continuing voltage augmentation revealed that the phenomenon of agglomerative adsorption of micron-nano dust became less prominent, and the micron-nano dust speckle remained stably adsorbed on the insulator surface. At −52 kV, an intriguing event occurred: the micron-nano dust speckle abruptly "exploded", creating a ring-shaped micron-nano dust halo above the speckle. This ring-shaped halo, similar to the dust speckle, would stably adhere to the epoxy resin insulator's surface if the voltage was not increased further. Notably, the "explosion" of the dust speckle to form the dust halo was a brief event, occurring immediately after reaching −52 kV and concluding rapidly. This "explosion" lacked components of a redox reaction, in a chemical sense, but represented a physical phenomenon of rapid expansion. Concurrently, the ground electrode diffusion adsorption phenomenon intensified, enlarging the diffusion area, and deepening in color.

Following the formation of the ring-shaped micron-nano dust halo, further voltage escalation resulted in a notable instantaneous adsorption of the residual micron-nano dust on the ground electrode. Most of these particles adhered near the recently formed ring-shaped micron-nano dust halo, leading to a significant deepening of the halo, as evident in Figure 14. Simultaneously, rapid diffusion occurred among the micron-nano dust on the ground electrode.

4.1.1. Formation and Contour Deepening of Micron-Nano Dust Halos at Different Initial Positions

Illustrated in Figure 9, the adsorption experiments involving 20 mg of 50 nm aluminum dust at the initial positions 147, 456, 159, and 147/369 exhibit the formation of a stable micron-nano dust speckle on the epoxy insulator surface when the voltage reaches −40 kV, −42 kV, −42 kV, and −44 kV. Upon further voltage increase to −60 kV, −52 kV, −60 kV, and −60 kV, the micron-nano dust speckle undergoes an "explosion", resulting in the creation of a ring-shaped micron-nano dust halo diagonally above the dust speckle. At −66 kV, −64 kV, −66 kV, and −70 kV, a noticeable one-time instantaneous adsorption occurs among the residual micron-nano dust on the ground electrode. The majority of these particles adhere to the vicinity of the recently formed annular micron-nano dust halo, significantly deepening it, as depicted in Figure 15.

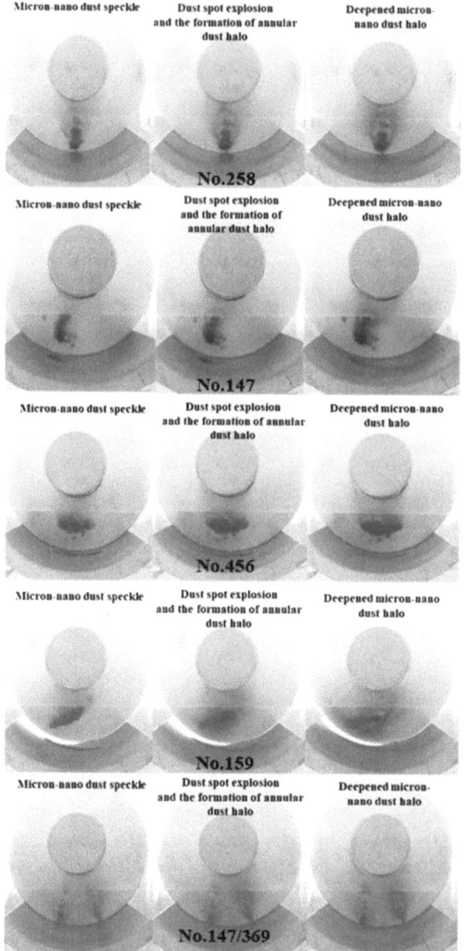

Figure 15. Special physical phenomena of dust speckles under different circumstances.

Despite variations in the morphology of the micron-nano dust speckles formed at distinct initial positions, the experimental phenomena involving the explosion formation of annular dust halos and the deepening of dust halo contours are consistently observed under DC voltage as the voltage increases.

4.1.2. Formation and Contour Deepening of Micron-Nano Copper Dust Halos

As depicted in Figure 16, in the adsorption experiment conducted with copper dust starting at position 258, when the voltage reaches −40 kV, a micron-nano copper dust speckle forms on the surface of the epoxy resin insulator. Upon further voltage increase to −60 kV, the micron-nano dust speckle undergoes an "explosion", resulting in the formation of dust halos on both sides of the speckle. Although not forming a closed circle, the dust halo exhibits an arc distribution, contributing to the creation of a ring-shaped micron-nano dust halo. At −72 kV, the contour of the dust halo deepens, as depicted in Figure 16.

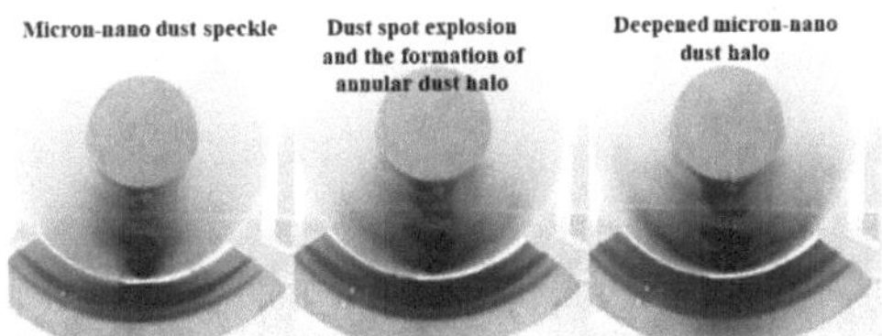

Figure 16. Copper dust halo formation and contour deepening at position 258.

4.1.3. Special Physical Phenomena Prior to Flashover along the Surface under AC Voltage

As illustrated in Figure 15, during the adsorption experiment involving 20 mg of 50 nm aluminum dust at the initial position 258 under externally applied AC voltage, upon reaching −42 kV, an unstable dust speckle agglomerates on the surface of the epoxy resin insulator. Subsequently, at 52 kV, the micron-nano dust speckle undergoes an "explosion", resulting in the formation of a larger surface area of micron-nano dust band, as shown in Figure 17. In contrast to DC voltage, AC voltage does not lead to the formation of a dust halo during the dust speckle explosion; instead, it directly completes an instantaneous adsorption, forming a larger area of agglomeration in the band dust speckle.

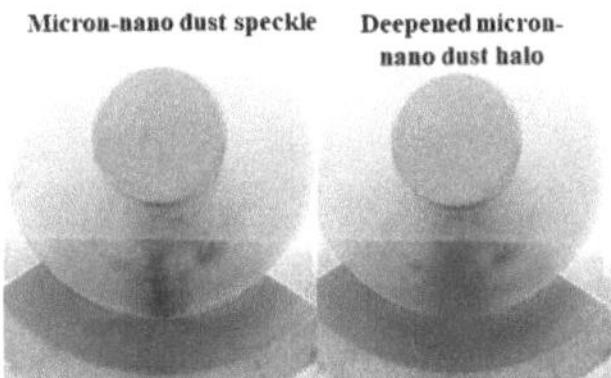

Figure 17. 20 mg, 50 nm aluminum dust band at position 258 under AC voltage.

In summary, when different materials of micron-nano dust are subjected to the same applied voltage, the experimental phenomenon of dust speckle explosion leading to the formation of a ring-shaped dust halo, followed by the deepening of the dust halo contour, varies. Copper dust and aluminum dust form similar dust speckles, but the shape of the dust halo resulting from the dust speckle "explosion" exhibits different morphologies, including arcs and rings. Nevertheless, it is undeniable that, as the voltage applied to the micron-nano dust speckle continues to rise, it gives rise to special physical phenomena distinct from the initial adsorption, thus creating necessary conditions for induced flashover along the surface.

4.2. Dominant Factors Contributing to the Occurrence of Dust Spot Explosions

In the pressurization process, the presence of stable charge micron-nano dust speckles on the surface of the epoxy resin insulator induces severe distortions in the gas-solid interface electric field. Dust particles experience intense electric field distortions within the gap of the dust speckle, leading to SF6 gas partial discharge. As a consequence, dust adheres to the insulator surface, causing an increase in internal temperature and promoting the agglomeration of dust particles within the speckle. This agglomeration results in a considerable amount of energy, causing excessive stress inside the dust speckle. The heightened stress leads to the outward splashing of dust particles, presenting a macroscopic "explosion phenomenon".

Simultaneously, micron-nano dust particles initially positioned on the ground electrode acquire the same charge induced by the high-voltage conductor. With the rising voltage, these similarly charged micron-nano dust particles lift to the insulator surface, forming micron-nano dust speckles. As the voltage continues to increase, Coulomb repulsion among the like-charged micron-nano dust particles within the speckle gradually intensifies. The lifting of dust and subsequent collisions disturb the equilibrium of the dust speckle, causing a small number of dust particles to bounce outward. This process is also characterized by a macroscopic "explosion phenomenon".

5. Physical Mechanism of Interfacial Adsorption of Micron-Nano Dust-Induced Flashover along the Epoxy Resin Surface

As depicted in Figure 15, the contour of the micron-nano dust halo experiences a pronounced deepening at −60 kV. With the ongoing increase in voltage, at −64 kV, the agglomerated micron-nano dust adhered to the surface of the epoxy resin insulator induces surface flashover. This flashover is characterized by the violent ablation of the epoxy resin surface. Furthermore, the current flowing along the surface during the flashover initiates a direct and intense reaction between the residual metal dust on the ground electrode and SF6, as illustrated in Figure 18.

Figure 18. Surface flashover induced by 20 mg of 50 nm aluminum powder at position 258.

5.1. Interfacial Adsorption of Micron-Nano Dust-Induced Flashover along the Surface

Despite differences in micron-nano dust speckle formation, the emergence of ring-shaped dust halo explosions, and the deepening of the dust halo contour, the flashover development consistently follows a path through the micron-nano dust speckle. The flashover current traverses the dust halo and dust speckle, sufficiently reacting at the epoxy resin surface to ultimately ignite residual dust on the ground electrode. In Figure 19, voltage values for each characteristic stage of micron-nano dust adsorption along the insulator surface leading to surface flashover are presented under diverse initial conditions.

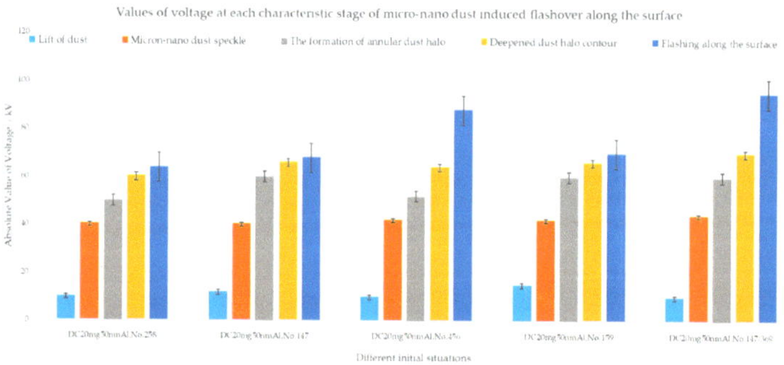

Figure 19. Values for the characteristic stages of micron-nano dust adsorption.

Upon analysis of Table 1, it becomes evident that the requisite condition for micron-nano dust to induce surface flashover under DC voltage is the formation of a micron-nano dust speckle and the subsequent "explosion" of the dust speckle. Unlike large-sized particles, DC voltage is more likely to cause insulation failure due to its unidirectional force, whereas the opposite holds true for micron-nano dust. Micron-nano dust poses a greater hazard under AC voltages, exhibiting significantly lower breakdown voltages than in DC environments under identical conditions.

Under DC voltage, the flashover voltage along the surface is greatly influenced by the initial position, initial mass, and dust material, with micron-nano dust reducing the breakdown voltage of the test system by up to 71%.

5.2. Mechanism of Action for Inducing Flashovers along the Surface

Micron-nano dust exhibits pronounced conductivity, and the agglomeration adsorption forming micron-nano dust speckles on the surface of epoxy resin insulators establishes a localized short-circuit region. This short-circuit region reduces the insulation distance between the high-voltage conductor and the grounding shell, making flashover more likely as the length of the dust speckle along the insulator bus direction increases, elongating the "short-circuit region" between the high and low voltage electrodes.

Microscopic analysis reveals that agglomeration adsorption along the epoxy resin surface of micron-nano dust spots leads to varying forms of existence and arrangements, causing different degrees of distortion in the surrounding electric field. In the discharge process, aluminum and copper micron-nano dust primarily exist in clusters, resulting in the most significant electric field distortion at the center sphere, reaching up to 4.5 times the original field strength. However, dust generated in the GIS/GIL cavity typically assumes irregular geometries due to mechanical wear, with dust having multiple spikes, leading to more severe local electric field distortion.

This distorted electric field exacerbates charge accumulation on the epoxy resin surface, further intensifying the surrounding electric field distortion. When the electric field distortion surpasses the local breakdown field strength of the gas between two particles, local discharge occurs, evolving into a flashover phenomenon along the epoxy resin surface.

On a more microscopic level, considering energy band structure and charge dissipation, external excitation and electric field distortion lead to the release of electrons from the valence band to the conduction band. Under the applied field strength, these liberated electrons initiate a flow of injection, generating new electrons through impacts on the surface. Metals, lacking a band gap, require less energy for electron jumps, making flash channel formation more likely. Due to external excitation reasons such as electric field distortion, thermal vibration occurs in the original thermal equilibrium state of the material, leading some electrons to break away from the conduction band of the metal dust. In the external field, the initial electrons hit resin surface, generating new electrons, and moving back and forth. The recombination forms a streamer, resulting in failure of the insulation between electrodes.

6. Conclusions

In this study, we constructed an experimental platform tailored for observing the adsorption-flashover of micron-nano dust, operational within GIS/GIL conditions. By doing so, we captured the motion behavior and adsorption process of micron-nano dust near epoxy resin insulators, shedding light on the unique physical phenomena during the adsorption process. The following conclusions were drawn:

(1) The adsorption states of micron-nano dust in the electric field encompass agglomerative adsorption along the insulator surface and diffusive adsorption along the ground electrode's direction. Key factors influencing motion behavior include the initial position of micron-nano dust, its mass, material, and the externally applied voltage.

(2) Agglomerated adsorption of micron-nano dust leads to the formation of dust speckles on the epoxy resin insulator's surface, playing a crucial role in inducing subsequent surface flashovers. As voltage increases, these dust spots undergo an "explosion", generating an annular dust halo and other unique physical phenomena distinct from lifting adsorption. These phenomena are considered necessary conditions for inducing surface flashovers.

(3) Agglomeration adsorption of micron-nano dust speckles on the epoxy resin insulator's surface creates a localized short-circuit region, reducing insulation distance between the high-voltage conductor and the grounding shell. This, combined with the drastic distortion of the gas-solid interface electric field induced by micron-nano dust speckles, intensifies local discharge between dust particles, making surface flashovers more likely. The presence of micron-nano dust resulted in a significant 71% reduction in the breakdown voltage of the test system.

Author Contributions: Conceptualization, N.X. and Q.L.; methodology, N.X. and B.L.; software, N.X. and Y.W.; validation, N.X. and N.Y.; formal analysis, N.X.; investigation, R.Y.; data curation, N.X. and F.Z.; writing—original draft preparation, N.X.; writing—review and editing, Q.L; supervision, N.X.; project administration, Q.L.; funding acquisition, N.X. All authors have read and agreed to the published version of the manuscript.

Funding: The work was funded in part by the National Natural Science Foundation of China (No. 52127812, 51929701) and The Fundamental Research Funds for the Central Universities (2023JC005).

Data Availability Statement: Data are contained within the article.

Conflicts of Interest: The authors declare no conflicts of interest.

References

1. Hu, Q.; Li, Q.; Liu, Z.; Xue, N.; Ren, H.; Haddad, M. Surface Flashover Induced by Metal Contaminants Adhered to Tri-Post Epoxy Insulators under Superimposed Direct and Lightning Impulse Voltages. *Polymers* **2022**, *14*, 1374. [CrossRef] [PubMed]
2. Zhou, H.; Ma, G.; Wang, C.; Wang, J.; Zhang, G.; Tu, Y.; Li, C. Review of charge accumulation on spacers of gas insulated equipment at DC stress. *CSEE J. Power Energy Syst.* **2021**, *7*, 510–529.
3. Magier, T.; Tenzer, M.; Koch, H. Direct current gas-insulated transmission lines. *IEEE Trans. Power Deliv.* **2018**, *33*, 440–446. [CrossRef]
4. Kumada, A.; Okabe, S. Charge distribution measurement on a truncated cone spacer under DC voltage. *IEEE Trans. Dielectr. Electr. Insul.* **2004**, *11*, 929–938. [CrossRef]
5. Hasegawa, T.; Yamaji, K.; Hatano, M.; Aoyagi, H.; Taniguchi, Y.; Kobayashi, A. DC dielectric characteristics and conception of insulation design for DC GIS. *IEEE Trans. Power Deliv.* **1996**, *11*, 1776–1782. [CrossRef]
6. Xue, N.; Yang, J.; Shen, D.; Xu, P.; Yang, K.; Zhuo, Z.; Zhang, L.; Zhang, J. The Location of Partial Discharge Sources inside Power Transformers Based on TDOA Database with UHF Sensors. *IEEE Access* **2019**, *7*, 146732–146744. [CrossRef]
7. Zhou, H.; Ma, G.; Li, C.; Shi, C.; Qin, S. Impact of temperature on surface charges accumulation on insulators in SF6 filled DC-GIL. *IEEE Trans. Dielectri CS Electr. Insul.* **2017**, *24*, 601–610. [CrossRef]
8. Qi, B.; Li, C.; Hao, Z.; Geng, B.; Xu, D.; Liu, S.; Deng, C. Surface discharge initiated by immobilized metallic particles attached to gas insulated substation insulators: Process and features. *IEEE Trans. Dielectr. Electr. Insul.* **2011**, *18*, 792–800. [CrossRef]
9. Wang, J.; Wang, J.; Hu, Q.; Chang, Y.; Liu, H.; Liang, R. Mechanism analysis of particle-triggered flashover in different gas dielectrics under DC superposition lightning impulse voltage. *IEEE Access* **2020**, *8*, 182888–182897. [CrossRef]
10. Ma, G.; Zhou, H.; Wang, Y.; Zhang, H.; Lu, S.; Tu, Y.; Wang, J.; Li, C. Flashover behavior of cone-type spacers with inhomogeneous temperature distribution in SF$_6$/N$_2$-filled DC-GIL under lightning impulse with DC voltage superimposed. *CSEE J. Power Energy Syst.* **2020**, *6*, 427–433.
11. Chang, Y.; Liu, Z.; Li, Q.; Xue, N.; Wang, J.; Manu, H. Capture mechanism and optimal design of microparticle traps in HVAC/HVDC gas insulated equipment. *IEEE Trans. Power Deliv.* **2022**, *37*, 4700–4710. [CrossRef]
12. Liang, R.; Hu, Q.; Liu, H.; Wang, J.; Li, Q. Research on Discharge Phenomenon Caused by Cross-Adsorption of Linear Insulating Fiber and Metal Dust under DC Voltage. *High Voltag.* **2020**, *7*, 269–278. [CrossRef]
13. Wang, J.; Li, Q.; Gong, Y.; Su, B.; Li, Z.; Liu, H.; Wang, J.; Ren, H. Enhancement of surface electrical performances of epoxy film by blending with fluorinated and crosslinking monomer. *Mater. Lett.* **2022**, *327*, 133036. [CrossRef]
14. Zhang, L.; Lu, S.; Li, C.; Cui, B.; Wu, Y. Movement and Discharge Characteristics of Micron-Scale Metal Dust in Gas Insulated Switchgear. *Trans. China Electrotech. Soc.* **2020**, *35*, 444–452.
15. Li, X.; Liu, W.; Xu, Y.; Cheng, W.; Bi, J. Surface charge accumulation and pre-flashover characteristics induced by metal particle on 1100kV GIL post insulator surface under AC voltage. *High Volt.* **2020**, *5*, 134–142.

Article

Research on Performance Evaluation of Polymeric Surfactant Cleaning Gel-Breaking Fluid (GBF) and Its Enhanced Oil Recovery (EOR) Effect

Yubin Liao [1],*, Jicheng Jin [2], Shenglin Du [2], Yufei Ren [2] and Qiang Li [1],*

[1] College of Construction Engineering, Jilin University, Changchun 130026, China
[2] No. 4 Oil Production Plant, PetroChina Qinghai Oilfield Company, Mangya 816400, China; jjcqh@petrochina.com.cn (J.J.); dslbyqh@petrochina.com.cn (S.D.); ryfqh@petrochina.com.cn (Y.R.)
* Correspondence: liaoyb16@mails.jlu.edu.cn (Y.L.); lqiang1982@jlu.edu.cn (Q.L.)

Abstract: Clean fracturing fluid has the characteristics of being environmentally friendly and causing little damage to reservoirs. Meanwhile, its backflow gel-breaking fluids (GBFs) can be reutilized as an oil displacement agent. This paper systematically evaluates the feasibility and EOR mechanism of a GBF based on a polymer surfactant as an oil displacement system for reutilization. A rotating interfacial tensiometer and contact angle measuring instrument were used to evaluate the performance of reducing the oil–water interfacial tension (IFT) and to change the rock wettability, respectively. Additionally, a homogeneous apparatus was used to prepare emulsions to evaluate GBF's emulsifying properties. Finally, core flooding experiments were used to evaluate the EOR effect of GBFs, and the influence rules and main controlling effects of various properties on the EOR were clarified. As the concentration of GBFs increases, the IFT first decreases to the lowest of 0.37 mN/m at 0.20 wt% and then increases and the contact angle of the rock wall decreases from 129° and stabilizes at 42°. Meanwhile, the emulsion droplet size gradually decreases and stabilizes with increases in GBF concentration, and the smallest particle size occurs when the concentration is 0.12–0.15 wt%. The limited adsorption area of the oil–water interface and the long molecular chain are the main reasons that limit the continued IFT reduction and emulsion stability. The oil displacement experiment shows that the concentration of GBF solution to obtain the best EOR effect is 0.15 wt%. At this concentration, the IFT reduction and the emulsification performance are not optimal. This shows that the IFT reduction performance, reservoir wettability change performance, and emulsification performance jointly determine the EOR effect of GBFs. In contrast, the emulsifying performance of GBFs is the main controlling factor for the EOR. Finally, the optimal application concentration of GBFs is 0.15–0.20 wt%, and the optimal injection volume is 0.5 PV.

Keywords: shale oil; clean fracturing fluids; gel-breaking fluids; reutilization; emulsion; EOR

Citation: Liao, Y.; Jin, J.; Du, S.; Ren, Y.; Li, Q. Research on Performance Evaluation of Polymeric Surfactant Cleaning Gel-Breaking Fluid (GBF) and Its Enhanced Oil Recovery (EOR) Effect. *Polymers* **2024**, *16*, 397. https://doi.org/10.3390/polym16030397

Academic Editor: Alberto Romero García

Received: 23 December 2023
Revised: 19 January 2024
Accepted: 22 January 2024
Published: 31 January 2024

1. Introduction

As conventional oil and gas resources enter the middle and late development stages, the development and utilization of unconventional oil and gas resources such as shale oil, tight oil, and heavy oil have become the main topics of scientific research and engineering construction [1–4]. The marine shale oil revolution in the United States has promoted the rapid growth of shale oil and gas at an average annual rate of 25% in the past 10 years [5]. However, the development of continental shale oil in China has been more difficult and is still in the development stage [6,7]. Continental shale is mainly formed in semi-deep-water to deep-water lacustrine sedimentary environments. The distribution area of the shale series is small, with diverse lithofacies types and poor structural environment stability [8,9]. Meanwhile, continental shale reservoirs are highly heterogeneous, with well-developed micro- and nano-pores and low development of organic pores, which leads to a complex distribution and small area of the "sweet areas" [10,11]. The above geological characteristics of continental shale oil have led to problems in shale oil development, such as the inability to inject and produce, rapid reduction in natural production capacity, and high precision requirements for production equipment. Hydraulic fracturing is a key production and injection technology for efficient exploitation of shale reservoirs [12,13]. It can connect organic pores through hydraulic fracturing fractures and use proppant to maintain the opening of fractures, greatly improving reservoir permeability and fluid mobility [14–16]. Conventional water-based gel-fracturing fluids (slick water, etc.,) have been widely used due to their low friction resistance and strong sand-carrying ability [17]. However, this type of polymer-based gel fracturing fluid produces a large amount of residue, which is an important source of pollution in oil field development as the flowback fluid is produced [18,19]. Developing a clean fracturing fluid system that is safe, environmentally friendly, and reusable while ensuring the basic performance of the fracturing fluid is key to promoting the efficient development of shale reservoirs.

The development and use of clean fracturing fluid originated at the end of the 20th century [20]. It is based on surfactant polymerization to form a viscoelastic surfactant gel as a thickened fracturing fluid system. This type of surfactant is generally a quaternary ammonium salt fatty acid surfactant, which exists as a spherical micelle in water [21–23]. When a salt solution is added to the surfactant solution, the spherical micelles further polymerize to form rod-shaped micelles (also called worm micelles), thereby significantly increasing the viscosity of the solution. When the micellar solution encounters oil or gas, it destroys the interaction between the surfactant and the salt and automatically breaks the gel [24,25]. The viscosity of the solution is significantly reduced to facilitate the backflow of the gel-breaking fluids. The cleaning fracturing fluid does not require the addition of chemical agents to form and break the gel, and the residual rate of the flowback solution is almost zero [26]. Therefore, clean fracturing fluid has the advantages of low reservoir damage, low environmental pollution, and single composition of the flowback solution and can be reutilized as an oil-displacement system [22]. Dantas et al. [27] conducted stable shear and oscillatory shear experiments to evaluate the rheological properties of anionic surfactant gel-breaking fluids. He pointed out that small changes in the surfactant will seriously affect the polymerization effect of the micelles and reveal its microscopic mechanisms. Legemah et al. [28] developed a new low-polymer-loaded fracturing fluid containing a boron cross-linking agent, which can significantly improve the sand-carrying capacity of the fracturing fluid while reducing the damage of polymer residues to reservoirs. Chieng et al. [29] improved the traditional single-chain viscoelastic surfactant and proposed a new type of high-temperature-resistant and shear-resistant thickening fracturing fluid. Li et al. [30] pointed out that a large amount of active substances in clean fracturing fluid will reduce the oil–water interfacial tension, change the wettability of the reservoir, and weaken the capillary driving pressure. However, the experimental results found that clean fracturing fluid has a good effect on improving oil recovery, which is attributed to the contribution of osmotic pressure. Meanwhile, Dai et al. also pointed out that osmotic pressure is an additional phenomenon that reduces interfacial tension and is the main controlling factor for gel-breaking fluids to improve oil recovery through studies on the adsorption rules of surfactants at the oil–water rock interface [31,32]. In addition, CO_2-responsive clean fracturing fluid also has good application prospects [33–35].

Additionally, a surfactant is an important chemical agent in tertiary oil recovery [36]. It is usually used in combination with polymers and microspheres or the form of a polymeric surfactant as an oil displacement agent [37–39]. It can improve oil recovery by reducing the capillary number, changing reservoir wettability, emulsifying crude oil, and modifying the reservoir profile [40–42]. The main component of gel-breaking fluids is the surfactant, which not only plays the role of imbibition and oil recovery during the fracturing backflow process but can also be reutilized as an oil displacement agent in the fracture and pore throat. Wang et al. [43] considered the coupling effects of wettability changes and interfacial tension reduction on surfactant adsorption and revealed the mechanism of wettability changes in different surfactants in carbonate and sandstone reservoirs. The change in reservoir wettability caused by surfactants is considered an important EOR mechanism when injecting liquid into low-permeability shale reservoirs [44–46]. Based on micropore-scale visualization experimental research, Yekeen et al. [47] explored the synergistic effect of nanoparticles and surfactants in reducing oil–water interfacial tension, changing rock wettability, and stabilizing emulsion. He pointed out that nano-surfactants show a good tendency to significantly reduce the oil–water interfacial tension and change the rock-wetting properties [48–52].

The above-mentioned research found that surfactant-based clean fracturing fluid systems are beneficial to replace traditional thickened gel fracturing fluids and can be reutilized as oil displacement systems after the fracturing fluids flow back. The current research has the following three problems: (1) the current clean fracturing fluid systems are all based on low-molecular surfactants, and there is a lack of research on clean fracturing fluid systems based on long-chain polymer surfactants; (2) at present, the synthesis of clean fracturing fluids only pursues the properties of ultra-low interfacial tension and ignores the main control effect of the emulsification performance of the gel-breaking fluid on the EOR; (3) most research on the EOR mechanisms of gel-breaking fluids are aimed at the imbibition process, lacking the EOR potential of reducing interfacial tension, wettability modification, and emulsification in its reutilized process.

Therefore, based on the above problems, this paper conducts experimental research on a clean fracturing fluid and its gel-breaking fluid based on polymer surfactants. The interfacial tension reduction performance, wettability improvement performance, and emulsification performance were evaluated, and the influence of concentration on the above properties was clarified. Finally, the relationship between the EOR and corresponding properties under different gel-breaking fluid concentrations is compared, and the EOR mechanism and the main control role of each property in its reutilized process are explained. Finally, the gel formation, gel breaking, and interfacial adsorption processes of cleaning fracture fluids are illustrated through illustrations. This paper highlights that under non-ultra-low interfacial tension, gel-breaking fluids can give full play to the main control role of emulsification in improving oil recovery and have good potential for oil displacement and reutilization.

2. Material and Method

2.1. Materials

Raw materials: Dichloromethane and ethanol used for core oil washing were purchased from KeLong Chemical Co., Ltd. (Chengdu, China); Inorganic salts such as NaCl, KCl, $CaCl_2$, etc., were purchased from Aladdin Biochemical Technology Co., Ltd. (Shanghai, China); Ethylene glycol monobutyl ether used for preparing the breaking fluids was purchased from Winson New Material Technology Co., Ltd. (Shanghai, China). Deionized water (DI water) was prepared by our lab using one comprehensive deionized water ultrapure water machine (WPL-EDI-DI-UP-10-UVF, Shanghai, China).

Clean fracturing fluids (CFF): CFF is mainly composed of a cationic viscoelastic polymer surfactant (VEPS), whose main component is long-chain tertiary amine (LCTA). By configuring aqueous solutions with LCTA and KCl at a mass ratio of 3 wt% and 4 wt%, respectively, the clean fracturing fluid CFF with good suspension properties was obtained, with a viscosity of approximately 120 cP at 70 °C.

Gel-breaking clean fracturing fluids (GBFs): For the GBFs, 1% kerosene was added to the CFF solution, which was stirred evenly at 25 °C for about 90 min until the gel broke. After the system was fully stable, we used a centrifuge to separate the kerosene to obtain the required gel-breaking fluid. The quality of the solution remained unchanged during the preparation process of the CFF and GBF, so the effective concentration of LCTA in the GBF solution was 3 wt%.

Solution: The degassed and dehydrated crude oil came from the Changqing Oilfield in China, and its viscosity and density at 70 °C were 1.34 cP and 0.734 g/cm^3, respectively.

The simulated formation water was prepared by sequentially adding inorganic salts to DI water according to the formula in Table 1 (the total salinity was 37,683 mg/L).

Table 1. The composition of the simulated formation water.

Iron	K^+, Na^+	Ca^{2+}	Mg^{2+}	CO_3^{2-}	HCO_3^-	SO_4^{2-}	Cl^-	Total
Concentration, mg/L	12,429.68	1110.17	637.88	57.12	1252.12	342.86	21,853.51	37,683

Cores: The cores were natural cylindrical cores obtained in the laboratory from the Changqing Oilfield (diameter was 2.5 cm, length was 5–10 cm, and permeability was within 5 mD). There was crude oil in the natural cores, which needed to be removed using ethanol and dichloromethane (DY-4 core rapid oil washing instrument, Yangzhou, China). The natural cores were placed in an oven to dry after oil washing for 3 days, and then the dry weight was recorded as m_1. The detailed parameters of the cores used in Section 2.2.4 are shown in Table 2.

Table 2. Basic core parameters.

Core	Pore Volume, mL	Diameter, cm	Length, cm	Permeability, mD	Volume of Oil, mL	Oil Saturation, %
C1	6.8	2.5	7.1	7.5	4.9	72.06
C2	7.1	2.5	7.2	8.3	5.2	73.24
C3	6.3	2.5	6.8	7.2	4.5	71.43
C4	6.5	2.5	6.9	7.5	4.7	72.31
C5	6.7	2.5	7.0	7.7	4.9	73.13
C6	6.7	2.5	7.1	7.8	4.7	70.15
C7	6.9	2.5	7.1	7.9	4.9	71.01

2.2. Methods

2.2.1. Evaluation of Interfacial Tension

Reducing the oil–water interfacial tension is a basic property of clean fracturing fluid, which can increase oil recovery by imbibition of the fracturing fluid matrix, making the crude oil easier to remove. A rotating interfacial tensiometer (Texas-500-C, 10^2–10^{-5} mN/m, Kono, Seattle, WA, USA) was used to measure the interfacial tension between CFF solution with different concentrations and dehydrated crude oil under 80 °C. To ensure the accuracy of the experimental results, a parallel control experiment was conducted for each group of IFT test experiments.

The specific test steps are: (1) Take the GBF solution prepared in the Materials section and dilute it with simulated formation water to concentrations of 0.02 wt%, 0.05 wt%, 0.08 wt%, 0.12 wt%, 0.15 wt%, and 0.20 wt%, 0.25 wt%, 0.30 wt%. (2) Use a needle to take a GBF solution and fill it into the interfacial tension meter test tube, being careful not to retain bubbles. (3) Use another special needle to absorb the crude oil and squeeze it into the solution in the tube. The main oil droplets (about 0.1 mL) should be of moderate size and should not be in contact with the wall of the test tube. (4) Turn on the temperature control system of the interfacial tension meter and wait for 2 h of constant temperature before testing. (5) Turn on the rotating interfacial tension meter, rotate the test tube at 3000 rpm, and use the image acquisition system to record the shape of the oil droplet in real time. (6) When the shape of the oil droplet does not change, take a photo to record the shape of the oil droplet and use the software analysis system to calculate the oil–water interfacial tension. (7) Stop rotating the interfacial tension meter, take out, clean, and dry the test tube. (8) Use GBF solutions of other concentrations and repeat steps (2) to (7) to obtain the oil–water interfacial tension under different GBF solution concentrations.

2.2.2. Evaluation of Wettability

The pore throats of shale reservoirs are small, and the lipophilic characteristics of the rock surface cause an oil film to be adsorbed on the surface, which not only reduces the oil-washing efficiency but also reduces the effective flow radius of the pore throats, exacerbating the difficulty of oil development. When the reservoir is oil wet, the contact angle of the solution is greater than 90°. Here, a static-contact goniometer (DSA100, Kruss, Hasvink, Germany) is used to test the effect of different concentrations of GBF solutions on rock wettability. The contact angle test software is equipped with an analysis system. Here, considering the contact angle range, the goniometric method is used to automatically measure and calculate the contact angle.

The specific test steps are: (1) Place several natural core sample slices and crude oil in a Petri dish at a volume ratio of 1:4 and age them for 1 day to obtain strongly lipophilic core slices. (2) Take the GBF solution prepared in the Materials section and dilute it with simulated formation water to concentrations ranging from 0.05 and 0.3 wt%. (3) Turn on the contact angle tester, and use a needle to absorb the GBF solution and drop it on the core slice (about 0.1 mL). (4) Place the core slice on the test bench, take pictures and record the images when the droplet shape does not change, and use the processing system to calculate the contact angle. (5) Replace the core slices and GBF solution and repeat step (3) and step (4) to obtain the contact angles of solutions with different GBF concentrations. As a baseline, the contact angle of DI water is 129.47°.

2.2.3. Evaluation of Emulsion Ability

The emulsifying property of GBF is key to its role in improving oil recovery. The evaluation of emulsifying properties usually includes emulsification strength and emulsion stability. The emulsion strength is evaluated by the emulsion droplet size distribution curve. The smaller the emulsion droplet size and the more uniform the distribution, the higher the emulsification strength. The stability of the emulsion is evaluated by the change curve of the liquid separation rate of the emulsion after standing. The faster and more complete the delamination, the worse stability. Here, a dispersion homogenizer (FJ300-S digital, Youyi, Shanghai, China) was used to prepare an emulsion of GBF solution and crude oil, and the particle size distribution was observed using a stereomicroscope (XTL-7000C, Caikon, Shanghai, China).

The specific experimental steps are: (1) Prepare 20 mL of GBF solution with a concentration of 0.05 wt%, 0.08 wt%, 0.12 wt%, 0.15 wt%, and 0.20 wt%. (2) Weigh 10 mL GBF solution and 10 mL crude oil, respectively, into a 30 mL beaker. (3) Use a homogenizer to mix the solution at a speed of 2000 rpm for 1 min and then pour it into a 25 mL test tube. (4) Place the test tube in an 80 °C incubator and record the amount of water precipitated in the test tube at certain intervals. We defined the ratio of the volume of precipitated water in the test tube recorded each time to the volume of added water (i.e., 10 mL) as the liquid drainage rate. Drawing the curve of the liquid drainage rate over time can clarify the stability of the emulsion. (5) Repeat step (3) to obtain the mixed emulsion, and draw a sample from the middle of the solution and drop it on a glass slide to observe its particle size using a stereomicroscope. (6) Replace the GBF solution and repeat steps (2) to (5) to evaluate the emulsion properties of solutions with different GBF concentrations.

2.2.4. Core Displacement Experiments

The reduction in interfacial tension, change in wettability, and emulsification ability of the GBF solution directly affect its final EOR effect. Meanwhile, the injection amount of the GBF solution will also affect its efficiency. Here, the influence of various parameters on the EOR effect of GBFs is evaluated through core flooding experiments. The specific experimental plan is shown in Table 3. The EOR effect of GBF flooding is the difference between the final oil recovery and the water flooding oil recovery. In order to ensure the accuracy of the experimental results, a parallel control experiment was conducted for each displacement experiment.

Table 3. Oil displacement experimental plan.

Number	Core	GBF Concentration, wt%	GBF Volume, PV	Injection Process	Note
1	C1	0.05	0.3		
2	C2	0.08	0.3		
3	C3	0.12	0.3	Water flooding to the water cut reaches 90%—GBF flooding to designed volume	Compare the influence of GBF concentration
4	C4	0.15	0.3		
5	C5	0.20	0.3		
6	C6	0.30	0.3		
7	C7	0.15	1.2		Compare the influence of GBF volume

The specific experimental steps are: (1) Put the core into the core holder and use a vacuum pump to evacuate the core end 1 for 4 h. (2) Saturate the simulated formation water. Connect end 2 of the core holder into a beaker containing simulated formation water and open the valve to self-absorb saturated water for 2 h. Record the volume change in the simulated formation water in the beaker before and after self-priming, which is the core pore volume V_1. Additionally, take out the saturated core and remove the surface water stains to weigh the wet weight m_2. Then, calculate the saturated water mass as m_1-m_2 and calculate the pore volume V_2 based on the simulated formation water density. Use V_1 and V_2 to correct the pore volume as $V = (V_1 + V_2)/2$ and calculate the porosity. (3) Absolute permeability test. Connect the experimental process according to Figure 1; use a constant speed pump to inject simulated formation water at a constant rate of 0.1 mL/min, and record the injection pressure. After recording the stable injection pressure, Darcy's law is used to calculate the core permeability. (4) Saturated crude oil. Use a constant speed pump to inject crude oil at a constant rate of 0.05 mL/min until no water is produced at the production end. Increase the injection rate to 0.2 mL/min and continue to inject 1 PV of crude oil. Stop the injection and place it in an 80 °C oven for aging for 3 days. Record the volume of produced water as the saturated oil volume and calculate the oil saturation S_o. (5) Carry out the oil displacement experiment according to the plan in Table 3 and record the injection pressure and liquid production throughout the process. Then, you calculate water cut (percentage of produced water volume to produced liquid volume) and oil recovery factor (percentage of produced oil volume to saturated oil volume) to compare their changes overtime. Keep the experimental temperature kept at 80 °C using one thermostat, and carry out the displacement experiments at a constant injection rate of 0.1 mL/min.

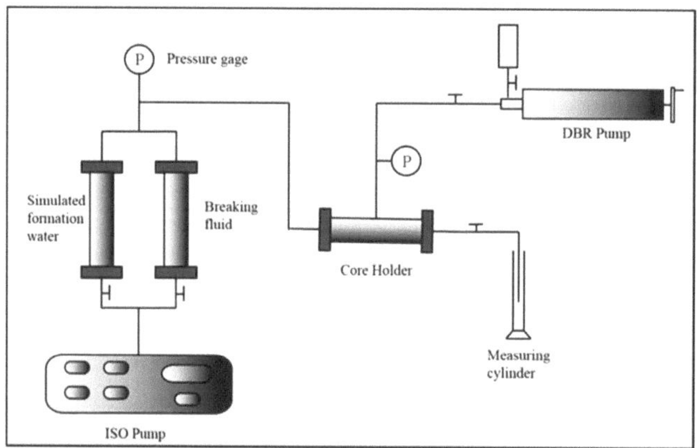

Figure 1. The flowchart of the displacement process.

3. Results and Discussion

3.1. IFT Reduction Ability of GBF

After fracturing in shale reservoirs, the clean fracturing fluid will break through contact with crude oil and become a GBF. The GBF can fully contact the crude oil and wall surface in the reservoir, and the decrease in IFT can lead to the decrease in the adhesion energy, making the crude oil more easily peeled off from the rock surface. The curve of IFT response to the GBF concentration is shown in Figure 2.

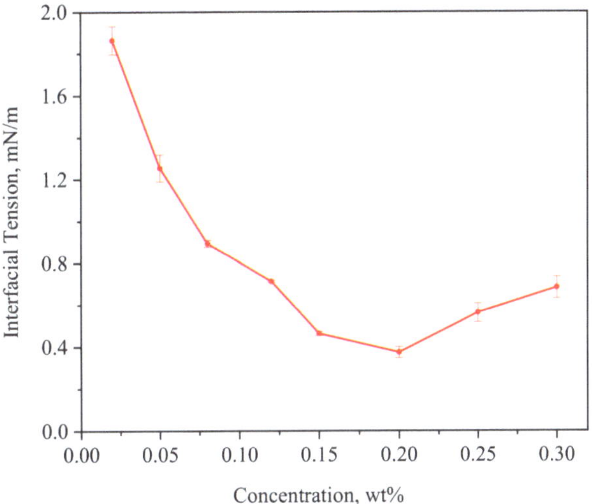

Figure 2. The steady-state IFT between crude oil and GBF with different concentration (the IFT between crude oil and DI water is 25.42 mN/m).

The steady-state IFT between crude oil and the GBF first decreases rapidly with the increase in GBF concentration, reaching the lowest value of 0.37 mN/m when the concentration is 0.2 wt%. Then, as the GBF concentration continues to increase, the IFT increases slowly. The CFF system used in this paper is a polymeric surfactant gel. The molecular chain of this surfactant is significantly larger than that of conventional surfactants, so its ability to reduce IFT is limited. This type of surfactant has a similar EOR mechanism to the polymeric surfactant and can achieve a good oil-washing effect at a relatively low IFT (the oil–water IFT test result is 34.26 mN/m). As the GBF concentration increases, there are two main reasons for the upward trend of IFT: (1) The oil–water interface area is limited, and continuing to increase the concentration of active agent molecules cannot increase the effective adsorption capacity. Meanwhile, the GBF is a long-chain molecule, which will reduce the adsorption area of active groups after adsorption at the interface. (2) GBF is a gel-breaking solution of CFF. When the concentration of the GBF gradually increases, it is easy to form a micelle gel again, thereby reducing the interfacial adsorption of the GBF.

In addition, it can be found that as the GBF concentration increases, the error of IFT first gradually decreases and then increases again. This is also related to the interface adsorption of GBF molecules. When the GBF concentration is low or high, the adsorption number of molecules at the interface will be affected by the solution preparation process, resulting in fluctuations in IFT test results.

3.2. Wettability Alteration Ability of GBF

The GBF solution can strip crude oil by reducing the oil–water IFT; in addition, the oil displacement efficiency is closely related to the wettability of the rock. The oil-wet surface causes the oil displacement efficiency to deteriorate, but the water-wet surface can increase the oil displacement efficiency. Therefore, the change in reservoir wettability is another main EOR mechanism of the GBF solution. The change curve of contact angle with the GBF solution concentration is shown in Figure 3.

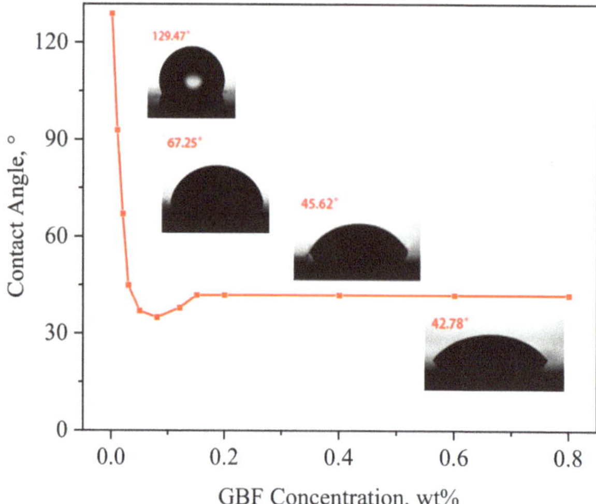

Figure 3. Variation curve of contact angle of oil-wet core thin sections with GBF concentration (the contact angle of DI water is 129.47°).

It can be seen from Figure 3 that the wetting angle decreases sharply with the increase in the GBF concentration within 0.1 wt% and then stays stable with the further increase in concentration. The reason is that GBF molecules are adsorbed on the oil-wet quartz surface through electrostatic attraction, Van Der Waals force, hydrogen bond, and hydrophobic action. The surfactant molecule hydrophobic group assembles directly with a solid surface and the hydrophilic group makes contact with an aqueous solution, which reduces the interfacial energy, and adhesion tension increases, s o the contact angle steeply falls, leading to enhanced water wettability on the surface of the quartz plate.

As the GBF concentration further increases, the adsorption capacity of the surfactant molecules increases, and the adsorption of surfactant molecules on the surface of quartz can cause a decrease in electronegativity. When the GBF concentration reaches a certain value, the interfacial energy and adhesion tension basically remain the same, so the contact angle tends to be stable at 42°. This suggests that the GBF solution can change the quartz surface from oil-wet to water-wet, which, combined with the reduction in oil–water IFT, greatly improves the oil-washing efficiency of shale reservoirs.

3.3. Emulsification Property of GBF

The GBF solution strips crude oil by reducing the IFT and changing rock wettability, which is the first step in improving oil recovery. The second step is to emulsify the stripped crude oil into an interfacial energy minimization system in which oil and water are evenly dispersed so that it can be easily carried and extracted in the form of emulsion droplets.

Figure 4 is the microscopic morphology of the droplet size distribution after crude oil is emulsified with GBF solutions of different concentrations. It can be found that as the concentration of the GBF solution increases, the emulsion droplet size first gradually decreases and then remains stable. It can be clearly observed that the granularity of the emulsion droplets has become weaker, which is the result of the high dispersion of the emulsion droplets. When the GBF solution concentration is 0.2 wt%, although the particle size of the emulsion droplets does not increase significantly, there are some large oil lumps, which have a serious impact on its stability.

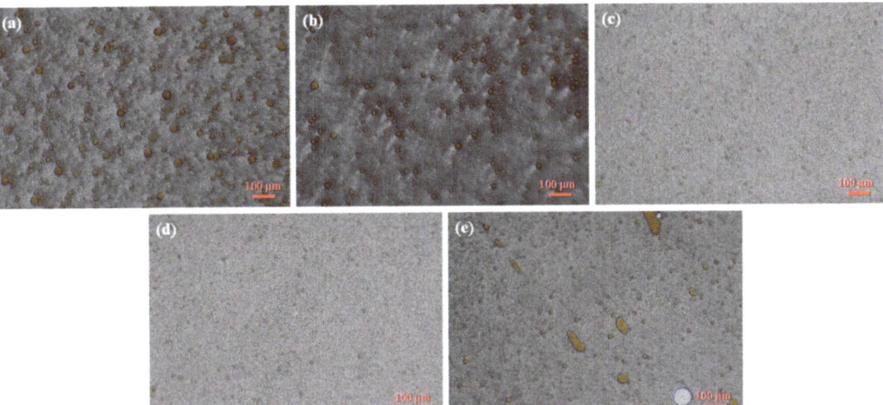

Figure 4. The droplet size of GBF emulsion with different concentrations. (**a**) GBF concentration is 0.05 wt%; (**b**) GBF concentration is 0.08 wt%; (**c**) GBF concentration is 0.12 wt%; (**d**) GBF concentration is 0.15 wt%; (**e**) GBF concentration is 0.20 wt%.

The liquid separation processes of the GBF emulsion with different concentrations are shown in Figure 5. It can be seen that the water drainage rate of the emulsion increases with the aging time and then stays stable. The liquid separation process of the GBF emulsion can last for 10–15 h, which is enough to illustrate its good stability. The changing trend of the final liquid drainage rate is that it first decreases and then increases with the increase in GBF concentration (as shown in the cyan dotted line in Figure 5). When the GBF concentration is 0.12 wt%, the liquid drainage rate is the lowest at 32%, and there is still a large amount of emulsion in the upper solution. When the GBF solution concentration is 0.20 wt%, the liquid drainage rate reaches a maximum of 85%. The change law of liquid drainage rate with the GBF concentration is consistent with its IFT reduction rule, and their mechanism of action is also the same. The decisive factor of the emulsion stability is the strength of the interface membrane. When the GBF concentration is small, fewer molecules are irregularly adsorbed on the interface and the membrane strength is weakening. So, the emulsion system is easy to demulsify and dehydrate with poor emulsion stability. When the GBF concentration is extremely large, the formation of micelles will damage the interface membrane, resulting in an increase in the water drainage rate.

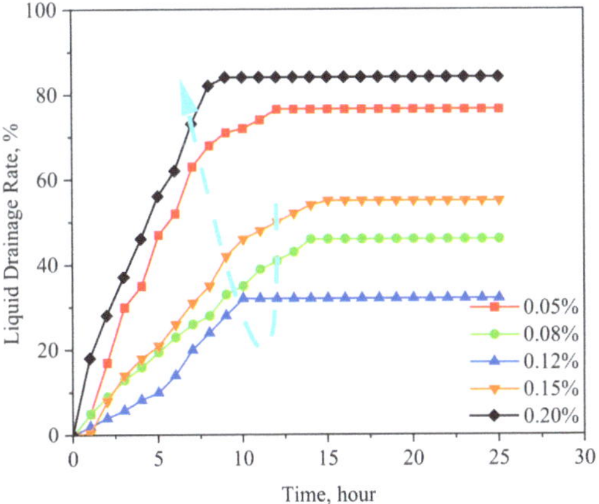

Figure 5. The water drainage rate of GBF change curves.

GBF strips crude oil and emulsifies it for extraction, which increases its oil-washing efficiency. Meanwhile, the emulsion can coalesce to form an oil band, and the oil band continues to encounter dispersed crude oil as it moves forward. The oil band formed by emulsified crude oil in the high permeability zone can modify the injection profile because of the Jamin effect. Thus, the GBF solution can also expand the swept volume until the oil band is taken out.

GBF is a potential EOR technology which has two EOR mechanisms: expanding the swept volume and improving the oil washing efficiency. Through the evaluation of the IFT reduction performance, wettability change performance, and emulsification performance, it can be determined that the optimal application concentration of GBF is 0.15 wt%. Its EOR efficiency will be evaluated in Section 3.4.

3.4. EOR Effect of GBF

Seven sets of GBF oil displacement experiments were carried out according to Table 3. The first six experiments of fixed injection volume were 0.3 PV to evaluate the impact of GBF concentration on the EOR effect. Meanwhile, the impact of IFT-reducing ability, wettability-changing ability, and emulsifying ability on the EOR at the corresponding concentration was evaluated. The seventh experiment fixed the GBF concentration at 0.15 wt% and performed a continuous GBF injection of 1.2 PV. During the period, the EOR at different injection volumes was intercepted to evaluate the impact of the GBF injection amount on the EOR effect.

The EOR and corresponding IFT, contact angle, and emulsion drainage stabilization time curves under six GBF concentrations are shown in Figure 6. The error bar of the EOR shows that the results of each experiment fluctuate within a certain range, but it can be found that this does not affect the change rule of EOR with GBF concentration. The main reasons for the error are the influence of experimental operations and the small differences in core parameters. The existence of the error bars proves the reliability of the experimental results and the objectivity of the rules. The EOR first increased significantly with the increase in GBF concentration and then showed a downward trend. The best EOR effect occurs when the GBF concentration is 0.15 wt%. Figure 6a shows that the concentration of the GBF solution with the lowest IFT is 0.20 wt%, and the EOR effect at this time begins to decrease. This is mainly because the stability of the GBF emulsion decreased significantly at this time, as shown in Figure 6c. Figure 6b shows that within the test concentration range, the contact angle is at a low value and tends to be stable, and its change pattern does not directly correspond to the EOR change rule. Figure 6c shows that the GBF concentration that obtains the best EOR is 0.15 wt%, which is inconsistent with the concentration corresponding to the lowest IFT and the lowest liquid drainage rate. This illustrates that the interfacial tension reduction performance and emulsification stability jointly determine the EOR effect of the GBF. However, the EOR when the liquid drainage rate is the lowest is closest to the optimal EOR. Overall, the increase and decrease in the EOR curve corresponds to the decrease and increase trend of the liquid drainage rate curve, which shows that among the three properties, the stability of the GBF emulsion is the main decisive factor controlling its EOR effect. Therefore, pursuing only ultra-low levels of IFT is not the correct direction for future oil displacement system construction.

The injection pressure will increase significantly after the GBF solution is injected. Although GBF can reduce the oil–water IFT to reduce pressure and increase injection, the emulsion formed at the same time will produce a Jamin effect at the pore throat and increase the injection pressure. The GBF injection pressure curves and corresponding EOR effects under different injection volumes are shown in Figure 7.

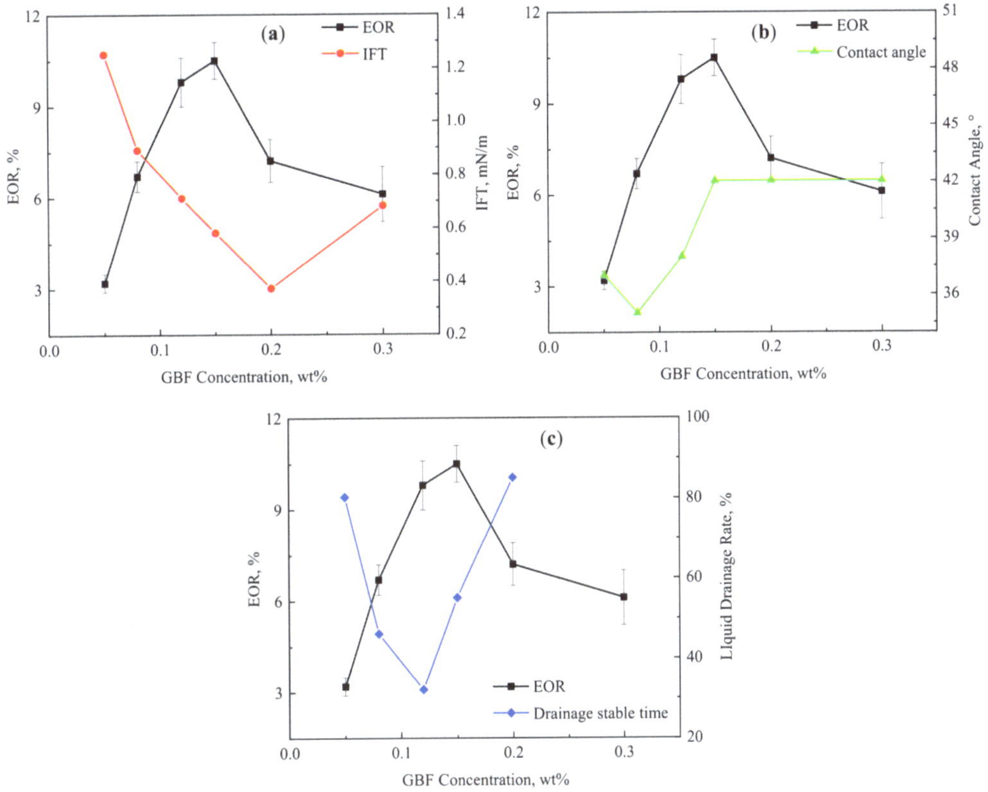

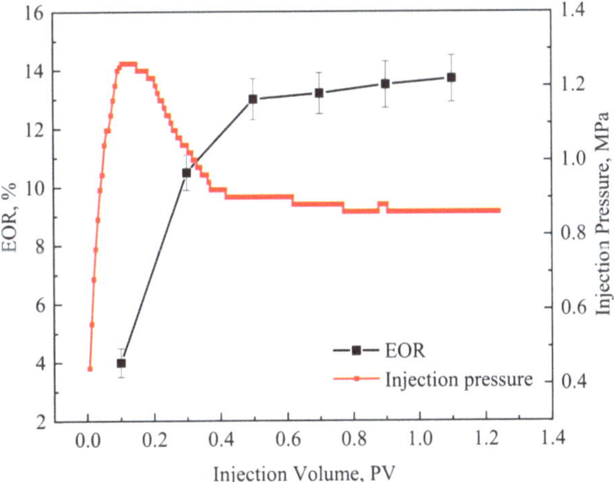

Figure 6. EOR and corresponding interface characteristic curves of GBF solutions at different concentrations. (**a**) IFT; (**b**) contact angle; (**c**) drainage stable time.

Figure 7. The relationship between the injection of slug and the growth rate of recovery.

Figure 7 shows that the injection pressure of the GBF stabilizes when the injection volume reaches 0.2 PV. At this time, the oil wall corresponding to the emulsion aggregation is produced. The EOR effect is significantly improved at this stage. After that, the injection pressure decreases rapidly and eventually levels off. It can be found that the moment when the EOR and the injection pressure become stable is when the injection volume is 0.5 PV, which shows that the increase in injection pressure is the main contribution to the improvement of the EOR effect. The experimental results also illustrate once again that the emulsifying performance of the GBF is the main controlling factor of its EOR efficiency.

Taking into account the various static properties and EOR effects of GBF, the optimal application concentration is 0.12–0.20 wt%, and the optimal injection volume is 0.5 PV. As an oil displacement agent for fracturing fluid reutilization, GBF is compared with a separate oil displacement system and other gel-breaking fluid, as shown in Table 4. It can be found that the ability of GBF to reduce interfacial tension is equivalent to that of a polymeric surfactant, but its viscosity is much lower, so the EOR effect is poor. Additionally, although the compound surfactant can achieve ultra-low interfacial tension and its EOR is higher than that of GBFs, its individual application is rare and the cost increases significantly. Although GBF does not have the advantages of ultra-low interfacial tension and high viscosity of surfactants and a polymeric surfactant, it takes into account the advantages of both. Compared with the EOR of about 10% of ordinary polymers, GBF has considerable EOR capabilities when reutilized as an oil displacement agent and has obvious economic advantages over surfactants.

Table 4. EOR comparison between GBF and other displacement agents.

Number	Agent	Concentration, wt%	IFT, mN/m	Volume, PV	Recovery, %	EOR, %
0	Gel-breaking fluid	0.15	0.46	0.5	58.32	13.00
1 [30]	Gel-breaking fluid	0.70	0.369	Imbibition	33.20	/
2 [31]	Gel-breaking fluid	2.00	0.14	Imbibition	40.00	/
3 [39]	Polymeric surfactant	1.50	~0.90	0.8	/	17.49
4 [53]	Compound surfactant	/	10^{-2}	3	/	13.65
5 [53]	Compound surfactant	/	10^{-3}	3	/	16.28
6 [54]	ASP (Alkali + Surfactant + Polymer)	S is 0.1–0.3	~10^{-2}	0.5–0.8	/	18–28

3.5. Mechanism of GBF Formed Process and EOR Effect

Gel-fracturing fluid processes such as CO_2 response [55], temperature response [56], ion response [29], etc., have the advantages of low viscosity during injection and easy gel breakage and are the development trend of clean fracturing fluid in the future. The clean gel fracturing fluid used in this paper is an ion-responsive polymer micelle gel. Current research has clarified its gel formation, gel breaking, and oil-displacement mechanisms, but there is a lack of comprehensive revelation and description of the above mechanisms. Based on the above experimental results and previous research experience, this paper summarizes the entire process mechanism of clean fracturing fluid, from gel formation to gel breaking and interface action to improve oil recovery.

A schematic diagram of the gel formation process of CFF, the gel breaking process to form GBF, and the adsorption mechanism of GBF at the interface, are shown in Figure 8. CFF is composed of a viscoelastic surfactant. In aqueous solutions, the nonpolar groups of these molecules tend to aggregate to avoid contact with water, eventually forming self-assembled spherical micelles. When the concentration of water anions increases, a large number of surfactant molecules further aggregate to form polymeric rod-like micelles, and the mutual entanglement between micelles greatly increases the viscoelasticity of the system. The viscoelasticity of the CFF solution is the basis for its significant sand-carrying performance. When organic matter or other hydrophobic substances (oil or gas) are dissolved in micelles, these rod-shaped micelles will be expanded and dispersed into smaller spherical micelles, resulting in reduced viscosity of the system. As the gel breaking has no breaking agent added, it will not cause any damage to the viscoelastic surfactant molecule and will provide a solid foundation for its further re-utilization.

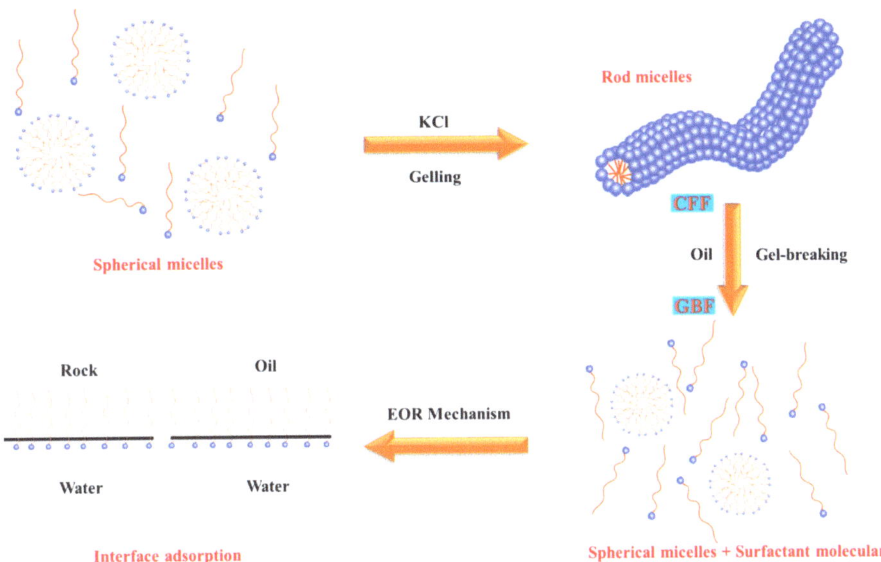

Figure 8. The process of clean fracturing fluid forming gels and breaking gels and the EOR mechanism.

In the application of the GBF flooding process, the surfactant molecules move toward and adsorb the oil–water interface and rock surface, effectively reducing the oil–water IFT and changing the wettability of the reservoir, which is of great significance to improving oil recovery.

4. Conclusions

This paper studies the reuse effect of gel-breaking fluids (GBFs) of a clean fracturing fluid system (CFF) based on polymer surfactants. The interfacial tension reduction performance, wettability change performance, and emulsification performance of GBFs were evaluated, and the main control effect of each factor on the EOR was clarified by comparing the EOR effect under various static properties, and the optimal application parameters were formed. Finally, the gelation and gel-breaking processes of CFF and the interfacial adsorption mechanism of the GBF solution were explained. The specific conclusions are as follows:

(1) GBFs can reduce the oil–water IFT to 10^{-1} mN/m. As the concentration of GBF increases, the oil–water IFT first decreases and then increases. The lowest IFT of 0.37 mN/m occurs when the GBF concentration is 0.20 wt%. The limited adsorption area of the oil–water interface and the long molecular chain are the main reasons that limit the continued IFT reduction.

(2) GBFs can effectively improve reservoir wettability. As the concentration of GBF solution increases, the contact angle of the rock wall decreases from 129° and stabilizes at 42°. Reducing the oil–water IFT and changing the wettability of the reservoir are the fundamental reasons why GBFs can effectively strip crude oil from shale reservoirs.

(3) GBFs have good emulsifying properties. As the concentration of GBF solution increases, the emulsion droplet size gradually decreases and stabilizes, with the smallest particle size at a concentration of 0.12–0.15 wt%. At a concentration of 0.20 wt%, a larger area of oil block will appear, which also corresponds to a significant reduction in emulsion stability.

(4) The optimal application concentration of GBFs is 0.12–0.20 wt%, and the optimal injection volume is 0.5 PV. The oil displacement experiment shows that the concentration of GBF solution to obtain the best EOR effect is 0.15 wt%. At this concentration, the IFT reduction and the emulsification performance are not optimal. This shows that the IFT reduction performance, reservoir wettability change performance, and emulsification performance jointly determine the EOR effect of GBFs. In contrast, the emulsifying performance of GBFs is the main controlling factor for the EOR.

(5) The experimental results of this paper prove that the BGF after breaking the CFF with a polymer surfactant as the main body has good EOR potential. It can effectively improve the oil washing efficiency and expand the swept volume at a lower dosage. Additionally, its main mechanism of action is emulsification rather than ultra-low interfacial tension, which also provides new thinking for the synthesis direction of oilfield chemicals.

Author Contributions: Y.L. conceived the presented idea. Y.L. completed the experiment and wrote the manuscript. Q.L. supervised this research and revised the manuscript. J.J., S.D. and Y.R. checked the experimental data. Q.L. supervised the findings of this work. All authors discussed the results and contributed to the final manuscript. All authors have read and agreed to the published version of the manuscript.

Funding: This work was funded by National Key R&D Projects (2019YFA0705501 and 2019YFA0705502) and the Science and Technology Innovation Team of Jilin University (2017TD-13).

Institutional Review Board Statement: Not applicable.

Data Availability Statement: Data are contained within the article.

Conflicts of Interest: Author Jicheng Jin, Shenglin Du, and Yufei Ren were employed by the PetroChina Qinghai Oilfield Company. The remaining authors declare that the research was conducted in the absence of any commercial or financial relationships that could be construed as a potential conflict of interest.

References

1. Burrows, L.C.; Haeri, F.; Cvetic, P.; Sanguinito, S.; Shi, F.; Tapriyal, D.; Goodman, A.; Enick, R.M. A literature review of CO_2, natural gas, and water-based fluids for enhanced oil recovery in unconventional reservoirs. *Energy Fuels* **2020**, *34*, 5331–5380. [CrossRef]

2. Malozyomov, B.V.; Martyushev, N.V.; Kukartsev, V.V.; Tynchenko, V.S.; Bukhtoyarov, V.V.; Wu, X.; Tyncheko, Y.A.; Kukartsev, V.A. Overview of Methods for Enhanced Oil Recovery from Conventional and Unconventional Reservoirs. *Energies* **2023**, *16*, 4907. [CrossRef]

3. Yang, H.; Lv, Z.; Zhang, M.; Jiang, J.; Xu, B.; Shen, J.; Jiang, H.; Kang, W. A novel active amphiphilic polymer for enhancing heavy oil recovery: Synthesis, characterization and mechanism. *J. Mol. Liq.* **2023**, *391*, 123210. [CrossRef]

4. Wang, W.; Xie, Q.; An, S.; Bakhshian, S.; Kang, Q.; Wang, H.; Xu, X.; Su, Y.; Cai, J.; Yuan, B. Pore-scale simulation of multiphase flow and reactive transport processes involved in geologic carbon sequestration. *Earth-Sci. Rev.* **2023**, *247*, 104602. [CrossRef]

5. Zou, C.; Ma, F.; Pan, S.; Zhang, X.; Wu, S.; Fu, G.; Wang, H.; Yang, Z. Formation and distribution potential of global shale oil and the developments of continental shale oil theory and technology in China. *Earth Sci. Front.* **2023**, *30*, 128.

6. Li, M.; Chen, X.; Wang, X. A Comparison of Geological Characteristics of the Main Continental Shale Oil in China and the US. *Lithosphere* **2021**, *2021*, 3377705. [CrossRef]

7. Wang, X.; Li, J.; Jiang, W.; Zhang, H.; Feng, Y.; Yang, Z. Characteristics, current exploration practices, and prospects of continental shale oil in China. *Adv. Geo-Energy Res.* **2022**, *6*, 454–459. [CrossRef]

8. He, W.; Zhu, R.; Cui, B.; Zhang, S.; Meng, Q.; Bai, B.; Feng, Z.; Lei, Z.; Wu, S.; He, K. The Geoscience Frontier of Gulong Shale Oil: Revealing the Role of Continental Shale from Oil Generation to Production. *Engineering* **2023**, *28*, 79–92. [CrossRef]

9. Sun, L.; He, W.; Feng, Z.; Zeng, H.; Jiang, H.; Pan, Z. Shale oil and gas generation process and pore fracture system evolution mechanisms of the Continental Gulong Shale, Songliao Basin, China. *Energy Fuels* **2022**, *36*, 6893–6905. [CrossRef]

10. Yang, Z.; Zou, C.; Wu, S.; Pan, S.; Wang, X.; Liu, H.; Jiang, W.; Li, J.; Li, Q.; Niu, X. Characteristics, types, and prospects of geological sweet sections in giant continental shale oil provinces in China. *J. Earth Sci.* **2022**, *33*, 1260–1277. [CrossRef]

11. Zou, C.-N.; Yang, Z.; Hou, L.-H.; Zhu, R.-K.; Cui, J.-W.; Wu, S.-T.; Lin, S.-H.; Guo, Q.-L.; Wang, S.-J.; Li, D.-H. Geological characteristics and "sweet area" evaluation for tight oil. *Pet. Sci.* **2015**, *12*, 606–617. [CrossRef]

12. Guo, T.; Zhang, S.; Qu, Z.; Zhou, T.; Xiao, Y.; Gao, J. Experimental study of hydraulic fracturing for shale by stimulated reservoir volume. *Fuel* **2014**, *128*, 373–380. [CrossRef]

13. Tan, P.; Jin, Y.; Han, K.; Hou, B.; Chen, M.; Guo, X.; Gao, J. Analysis of hydraulic fracture initiation and vertical propagation behavior in laminated shale formation. *Fuel* **2017**, *206*, 482–493. [CrossRef]

14. Cao, R.; Fang, S.; Jia, P.; Cheng, L.; Rao, X. An efficient embedded discrete-fracture model for 2D anisotropic reservoir simulation. *J. Pet. Sci. Eng.* **2019**, *174*, 115–130. [CrossRef]

15. Rao, X.; Cheng, L.; Cao, R.; Jia, P.; Liu, H.; Du, X. A modified projection-based embedded discrete fracture model (pEDFM) for practical and accurate numerical simulation of fractured reservoir. *J. Pet. Sci. Eng.* **2020**, *187*, 106852. [CrossRef]

16. Zhang, H.; Chen, J.; Zhao, Z.; Qiang, J. Hydraulic fracture network propagation in a naturally fractured shale reservoir based on the "well factory" model. *Comput. Geotech.* **2023**, *153*, 105103. [CrossRef]

17. Montgomery, C. Fracturing Fluids. In Proceedings of the ISRM International Conference for Effective and Sustainable Hydraulic Fracturing, ISRM, Brisbane, Australia, 20–22 May 2013; p. ISRM-ICHF-2013-035.

18. Huang, Q.; Liu, S.; Cheng, W.; Wang, G. Fracture permeability damage and recovery behaviors with fracturing fluid treatment of coal: An experimental study. *Fuel* **2020**, *282*, 118809. [CrossRef]

19. Wang, J.; Huang, Y.; Zhang, Y.; Zhou, F.; Yao, E.; Wang, R. Study of fracturing fluid on gel breaking performance and damage to fracture conductivity. *J. Pet. Sci. Eng.* **2020**, *193*, 107443. [CrossRef]

20. Guangyu, Y.; Jirui, H.; Huan, L. Current situation and development trend with respect to research and application of clean fracturing fluid. *China Surfactant Deterg. Cosmet.* **2012**, *42*, 288–292.

21. Wang, C.; Zhou, G.; Jiang, W.; Niu, C.; Xue, Y. Preparation and performance analysis of bisamido-based cationic surfactant fracturing fluid for coal seam water injection. *J. Mol. Liq.* **2021**, *332*, 115806. [CrossRef]

22. Yang, X.; Mao, J.; Zhang, H.; Zhang, Z.; Zhao, J. Reutilization of thickener from fracturing flowback fluid based on Gemini cationic surfactant. *Fuel* **2019**, *235*, 670–676. [CrossRef]

23. Zhao, G.; Yan, Z.; Qian, F.; Sun, H.; Lu, X.; Fan, H. Molecular simulation study on the rheological properties of a pH-responsive clean fracturing fluid system. *Fuel* **2019**, *253*, 677–684. [CrossRef]

24. Kang, W.; Mushi, S.J.; Yang, H.; Wang, P.; Hou, X. Development of smart viscoelastic surfactants and its applications in fracturing fluid: A review. *J. Pet. Sci. Eng.* **2020**, *190*, 107107. [CrossRef]

25. Yang, Y.; Zhang, H.; Wang, H.; Zhang, J.; Guo, Y.; Wei, B.; Wen, Y. Pseudo-interpenetrating network viscoelastic surfactant fracturing fluid formed by surface-modified cellulose nanofibril and wormlike micelles. *J. Pet. Sci. Eng.* **2022**, *208*, 109608. [CrossRef]

26. Huang, Q.; Liu, S.; Wu, B.; Wang, G.; Li, G.; Guo, Z. Role of VES-based fracturing fluid on gas sorption and diffusion of coal: An experimental study of Illinois basin coal. *Process Saf. Environ. Prot.* **2021**, *148*, 1243–1253. [CrossRef]

27. Dantas, T.C.; Santanna, V.C.; Neto, A.D.; Neto, E.B.; Moura, M.A. Rheological properties of a new surfactant-based fracturing gel. *Colloids Surf. A: Physicochem. Eng. Asp.* **2003**, *225*, 129–135. [CrossRef]

28. Legemah, M.; Guerin, M.; Sun, H.; Qu, Q. Novel high-efficiency boron crosslinkers for low-polymer-loading fracturing fluids. *SPE J.* **2014**, *19*, 737–743. [CrossRef]

29. Chieng, Z.; Mohyaldinn, M.E.; Hassan, A.M.; Bruining, H. Experimental investigation and performance evaluation of modified viscoelastic surfactant (VES) as a new thickening fracturing fluid. *Polymers* **2020**, *12*, 1470. [CrossRef] [PubMed]

30. Li, L.; Sun, Y.; Li, Y.; Wang, R.; Chen, J.; Wu, Y.; Dai, C. Interface properties evolution and imbibition mechanism of gel breaking fluid of clean fracturing fluid. *J. Mol. Liq.* **2022**, *359*, 118952. [CrossRef]

31. Zhao, M.; Liu, S.; Gao, Z.; Wu, Y.; Dai, C. The spontaneous imbibition mechanisms for enhanced oil recovery by gel breaking fluid of clean fracturing fluid. *Colloids Surf. A: Physicochem. Eng. Asp.* **2022**, *650*, 129568. [CrossRef]

32. Zhao, M.; Yan, X.; Cheng, Y.; Yan, R.; Dai, C. Study on the Imbibition Performance and Mechanism of a Fracturing Fluid and Its Gel Breaking Liquid. *Energy Fuels* **2022**, *36*, 13028–13036. [CrossRef]

33. Chen, X.; Li, Y.; Sun, X.; Liu, Z.; Liu, J.; Liu, S. Investigation of Polymer-Assisted CO_2 Flooding to Enhance Oil Recovery in Low-Permeability Reservoirs. *Polymers* **2023**, *15*, 3886. [CrossRef]

34. Xie, Q.; Wang, W.; Su, Y.; Wang, H.; Zhang, Z.; Yan, W. Pore-scale study of calcite dissolution during CO_2-saturated brine injection for sequestration in carbonate aquifers. *Gas Sci. Eng.* **2023**, *114*, 204978. [CrossRef]

35. Du, M.; Sun, X.; Dai, C.; Li, H.; Wang, T.; Xu, Z.; Zhao, M.; Guan, B.; Liu, P. Laboratory experiment on a toluene-polydimethyl silicone thickened supercritical carbon dioxide fracturing fluid. *J. Pet. Sci. Eng.* **2018**, *166*, 369–374. [CrossRef]

36. Isaac, O.T.; Pu, H.; Oni, B.A.; Samson, F.A. Surfactants employed in conventional and unconventional reservoirs for enhanced oil recovery—A review. *Energy Rep.* **2022**, *8*, 2806–2830. [CrossRef]

37. Chen, X.; Li, Y.; Liu, Z.; Li, X.; Zhang, J.; Zhang, H. Core-and pore-scale investigation on the migration and plugging of polymer microspheres in a heterogeneous porous media. *J. Pet. Sci. Eng.* **2020**, *195*, 107636. [CrossRef]

38. Chen, X.; Li, Y.; Liu, Z.; Tang, Y.; Sui, M. Visualized investigation of the immiscible displacement: Influencing factors, improved method, and EOR effect. *Fuel* **2023**, *331*, 125841. [CrossRef]

39. Chen, X.; Li, Y.-Q.; Liu, Z.-Y.; Gao, W.-B.; Sui, M.-Y. Experimental investigation on the enhanced oil recovery efficiency of polymeric surfactant: Matching relationship with core and emulsification ability. *Pet. Sci.* **2023**, *20*, 619–635. [CrossRef]

40. Chen, X.; Li, Y.; Liu, Z.; Zhang, J.; Chen, C.; Ma, M. Investigation on matching relationship and plugging mechanism of self-adaptive micro-gel (SMG) as a profile control and oil displacement agent. *Powder Technol.* **2020**, *364*, 774–784. [CrossRef]

41. Chen, X.; Li, Y.; Liu, Z.; Zhang, J.; Li, X. Experimental and theoretical investigation of the migration and plugging of the particle in porous media based on elastic properties. *Fuel* **2023**, *332*, 126224. [CrossRef]

42. Yang, H.; Lv, Z.; Li, Z.; Guo, B.; Zhao, J.; Xu, Y.; Xu, W.; Kang, W. Laboratory evaluation of a controllable self-degradable temporary plugging agent in fractured reservoir. *Phys. Fluids* **2023**, *35*, 083314. [CrossRef]

43. Wang, Y.; Xu, H.; Yu, W.; Bai, B.; Song, X.; Zhang, J. Surfactant induced reservoir wettability alteration: Recent theoretical and experimental advances in enhanced oil recovery. *Pet. Sci.* **2011**, *8*, 463–476. [CrossRef]

44. Giraldo, J.; Benjumea, P.; Lopera, S.; Cortés, F.B.; Ruiz, M.A. Wettability alteration of sandstone cores by alumina-based nanofluids. *Energy Fuels* **2013**, *27*, 3659–3665. [CrossRef]

45. Hou, J.; Sun, L. Synergistic effect of nanofluids and surfactants on heavy oil recovery and oil-wet calcite wettability. *Nanomaterials* **2021**, *11*, 1849. [CrossRef]

46. Karimi, A.; Fakhroueian, Z.; Bahramian, A.; Pour Khiabani, N.; Darabad, J.B.; Azin, R.; Arya, S. Wettability alteration in carbonates using zirconium oxide nanofluids: EOR implications. *Energy Fuels* **2012**, *26*, 1028–1036. [CrossRef]

47. Yekeen, N.; Padmanabhan, E.; Syed, A.H.; Sevoo, T.; Kanesen, K. Synergistic influence of nanoparticles and surfactants on interfacial tension reduction, wettability alteration and stabilization of oil-in-water emulsion. *J. Pet. Sci. Eng.* **2020**, *186*, 106779. [CrossRef]

48. Keykhosravi, A.; Simjoo, M. Enhancement of capillary imbition by Gamma-Alumina nanoparticles in carbonate rocks: Underlying mechanisms and scaling analysis. *J. Pet. Sci. Eng.* **2020**, *187*, 106802. [CrossRef]

49. Kuang, W.; Saraji, S.; Piri, M. A systematic experimental investigation on the synergistic effects of aqueous nanofluids on interfacial properties and their implications for enhanced oil recovery. *Fuel* **2018**, *220*, 849–870. [CrossRef]

50. Sun, Y.-P.; Xin, Y.; Lyu, F.-T.; Dai, C.-L. Experimental study on the mechanism of adsorption-improved imbibition in oil-wet tight sandstone by a nonionic surfactant for enhanced oil recovery. *Pet. Sci.* **2021**, *18*, 1115–1126. [CrossRef]

51. Tian, W.; Wu, K.; Gao, Y.; Chen, Z.; Gao, Y.; Li, J. A critical review of enhanced oil recovery by imbibition: Theory and practice. *Energy Fuels* **2021**, *35*, 5643–5670. [CrossRef]

52. Xu, D.; Li, Z.; Bai, B.; Chen, X.; Wu, H.; Hou, J.; Kang, W. A systematic research on spontaneous imbibition of surfactant solutions for low permeability sandstone reservoirs. *J. Pet. Sci. Eng.* **2021**, *206*, 109003. [CrossRef]

53. Yuan, C.; Pu, W.; Wang, X.; Sun, L.; Zhang, Y.; Cheng, S. Effects of interfacial tension, emulsification, and surfactant concentration on oil recovery in surfactant flooding process for high temperature and high salinity reservoirs. *Energy Fuels* **2015**, *29*, 6165–6176. [CrossRef]

54. Guo, H.; Li, Y.; Wang, F.; Gu, Y. Comparison of strong-alkali and weak-alkali ASP-flooding field tests in Daqing oil field. *SPE Prod. Oper.* **2018**, *33*, 353–362. [CrossRef]

55. Musevic, I.; Skarabot, M.; Tkalec, U.; Ravnik, M.; Zumer, S. Two-dimensional nematic colloidal crystals self-assembled by topological defects. *Science* **2006**, *313*, 954–958. [CrossRef]

56. Davies, T.; Ketner, A.; Raghavan, S. Self-assembly of surfactant vesicles that transform into viscoelastic wormlike micelles upon heating. *J. Am. Chem. Soc.* **2006**, *128*, 6669–6675. [CrossRef]

Article

Preparation and Performance Evaluation of Amphiphilic Polymers for Enhanced Heavy Oil Recovery

Dongtao Fei [1], Jixiang Guo [1,*], Ruiying Xiong [2], Xiaojun Zhang [1], Chuanhong Kang [1] and Wyclif Kiyingi [1]

[1] Unconventional Petroleum Research Institute, China University of Petroleum, Beijing 102249, China; fdt305@163.com (D.F.); zxjhuagong@163.com (X.Z.); k15616309001@163.com (C.K.); kiyingiwyclif@outlook.com (W.K.)

[2] College of Petroleum Engineering, China University of Petroleum, Beijing 102249, China; ryxiong08@163.com

* Correspondence: guojx003@126.com

Abstract: The continuous growth in global energy and chemical raw material demand has drawn significant attention to the development of heavy oil resources. A primary challenge in heavy oil extraction lies in reducing crude oil viscosity. Alkali–surfactant–polymer (ASP) flooding technology has emerged as an effective method for enhancing heavy oil recovery. However, the chromatographic separation of chemical agents presents a formidable obstacle in heavy oil extraction. To address this challenge, we utilized a free radical polymerization method, employing acrylamide, 2-acrylamido-2-methylpropane sulfonic acid, lauryl acrylate, and benzyl acrylate as raw materials. This approach led to the synthesis of a multifunctional amphiphilic polymer known as PAALB, which we applied to the extraction of heavy oil. The structure of PAALB was meticulously characterized using techniques such as infrared spectroscopy and Nuclear Magnetic Resonance Spectroscopy. To assess the effectiveness of PAALB in reducing heavy oil viscosity and enhancing oil recovery, we conducted a series of tests, including contact angle measurements, interfacial tension assessments, self-emulsification experiments, critical association concentration tests, and sand-packed tube flooding experiments. The research findings indicate that PAALB can reduce oil–water displacement, reduce heavy oil viscosity, and improve swept volume upon injection into the formation. A solution of 5000 mg/L PAALB reduced the contact angle of water droplets on the core surface from 106.55° to 34.95°, shifting the core surface from oil-wet to water-wet, thereby enabling oil–water displacement. Moreover, A solution of 10,000 mg/L PAALB reduced the oil–water interfacial tension to 3.32×10^{-4} mN/m, reaching an ultra-low interfacial tension level, thereby inducing spontaneous emulsification of heavy oil within the formation. Under the condition of an oil–water ratio of 7:3, a solution of 10,000 mg/L PAALB can reduce the viscosity of heavy oil from 14,315 mPa·s to 201 mPa·s via the glass bottle inversion method, with a viscosity reduction rate of 98.60%. In sand-packed tube flooding experiments, under the injection volume of 1.5 PV, PAALB increased the recovery rate by 25.63% compared to traditional hydrolyzed polyacrylamide (HPAM) polymer. The insights derived from this research on amphiphilic polymers hold significant reference value for the development and optimization of chemical flooding strategies aimed at enhancing heavy oil recovery.

Keywords: amphiphilic polymer; hydrophobic association; self-emulsification; heavy oil viscosity reduction; enhanced oil recovery

Citation: Fei, D.; Guo, J.; Xiong, R.; Zhang, X.; Kang, C.; Kiyingi, W. Preparation and Performance Evaluation of Amphiphilic Polymers for Enhanced Heavy Oil Recovery. *Polymers* **2023**, *15*, 4606. https://doi.org/10.3390/polym15234606

Academic Editor: Stanislav Rangelov

Received: 16 October 2023
Revised: 21 November 2023
Accepted: 22 November 2023
Published: 2 December 2023

1. Introduction

Petroleum is a vital energy source and chemical feedstock, constituting a global reserve of approximately 9–11 trillion barrels, with heavy oil and bitumen resources accounting for over 70% of the total reserves. Given the escalating demand for energy and chemical feedstocks, the development of heavy oil resources has garnered significant attention [1–3]. Heavy oil extraction is challenging due to its high viscosity and limited mobility, making viscosity reduction within the reservoir crucial for successful heavy oil recovery [4,5].

Chemical flooding technology has been proven to be an effective method to enhance crude oil recovery. This approach involves the application of alkali, surfactants, polymers, and other chemicals to reduce heavy oil viscosity, increase flooding fluid viscosity, improve the oil–water mobility ratio and expand the sweep volume of oil displacement, ultimately enhancing heavy oil recovery [6–10]. However, this complex system often experiences chromatographic separation within the reservoir, hindering effective oil recovery. In recent years, scholars have begun investigating the application of amphiphilic polymers in heavy oil recovery. These polymers, featuring both hydrophilic and hydrophobic groups, facilitate efficient emulsification of heavy oil, reducing the viscosity of the oil phase while increasing that of the water phase. This, in turn, enhances the sweep efficiency of oil displacement without suffering from chromatographic separation. Thus, amphiphilic polymers hold substantial promise for heavy oil recovery [11–16].

Li [16] synthesized a series of water-soluble long branched-chain amphiphilic copolymers, AAGASs, and applied in the enhanced oil recovery of heavy oil. Hydrophobically associating water-soluble copolymers (AAGs) with an epoxy group were first synthesized via the free-radical copolymerization of acrylamide and sodium 2-acrylamido-2-methylpropanesulfonic acid with glycidyl methacrylate. An amino-terminated amphiphilic copolymer (AS-N) was also prepared with acrylamide and sodium 4-styrenesulfonate via chain transfer polymerization. AAGASs were then obtained via the chain-extending reaction between the epoxy groups of the AAG and the amino group of the AS-N. The results revealed that AAGAS exhibited unique association properties in solution, displaying excellent surface and interfacial activity. Consequently, AAGAS solutions were capable of both thickening water and transforming high-viscosity heavy oil into low-viscosity oil-in-water emulsions. Notably, they achieved an impressive 96.8% reduction in heavy oil viscosity at a concentration of 1000 mg/L AAGAS-3.

Qin [17] prepared a dendrimer amphiphilic polymer (A-D-HPAM) via the aqueous solution polymerization of acrylic acid, acrylamide, 2-acrylamido-2-methylpropanesulfonic acid, hydrophobic monomer of cetyl dimethyl allyl ammonium chloride, and skeleton monomer. It exhibited experimentally proven superior polymer performance. Compared to hydrolyzed polyacrylamide and hydrolyzed polyacrylamide–sodium dodecyl benzene sulfonate solutions, A-D-PAM notably increased oil recovery by 15.1%. Chen [18] synthesized the amphiphilic polymer PTVR using acrylamide, sodium p-styrenesulfonate, lauryl methacrylate, and methyl-2-urea-4[1H]-pyrimidinone. The results highlighted the exceptional performance of PTVR, which was attributed to the synergistic effect of these monomers. Methyl-2-urea-4[1H]-pyrimidinone is a quadruple hydrogen bond monomer, which effectively disrupts the hydrogen bond interactions within asphaltene or resin aggregates. On the other hand, the long alkyl chain lauryl methacrylate improved the polymer's affinity with heavy oil, facilitating the emulsification of heavy oil.

Liu [19] synthesized an amphiphilic polymer, AD-1, via free radical polymerization using acrylamide, acrylamide, 2-acrylamido-2-methyl-1-propane sulfonic acid, and self-tailored monomers with long hydrophobic chains and benzene rings, and conducted an investigation into amphiphilic polymer AD-1's role in enhancing heavy oil recovery via hot water flooding. The findings demonstrated that AD-1 had the capacity to form stable aggregates and a three-dimensional network in solution, exhibiting excellent interfacial performance. Even at low concentrations, AD-1 could create dynamic water-in-oil emulsions, facilitating both emulsification and rapid demulsification. Yang [20] synthesized a betaine-type binary amphiphilic polymer, PAMA, from acrylamide and association monomer made in a laboratory. The study elucidated the mechanism by which varying hydrophobic group content influenced PAMA's performance. Experimental findings demonstrated that higher hydrophobic group content improved PAMA's ability to reduce surface tension and foster associations among hydrophobic groups. Increased hydrophobic group content enhanced the polymer's viscosity, thermal stability, viscoelastic properties, and salt-thickening behavior.

Tian [21] conducted a comparative analysis of amphiphilic polymers featuring mono-branched and multi-branched hydrophobic groups. The research showed that multi-branched amphiphilic polymers had a dual effect, increasing both the viscosity and interfacial viscoelasticity of the dispersion medium, resulting in more stable emulsions. Zhu [22] examined amphiphilic polymers with both single-tail and double-tail hydrophobic groups to understand their impact on solution properties. The findings indicated that polymers with double-tail hydrophobic groups exhibited a lower critical association concentration (CAC). This led to stronger hydrophobic associations and, consequently, a higher apparent viscosity contribution rate when polymer concentrations exceeded the CAC. Moreover, amphiphilic polymers with double-tail hydrophobic groups displayed improved temperature resistance, salt resistance, and mechanical shear resistance at equivalent polymer solution concentrations. Zhi [23] employed amphiphilic polymer as a heavy oil activator and investigated the changes in residual oil distribution following activator displacement, along with its performance in heavy oil activation. The results demonstrated that the activator could reduce oil–water interfacial tension, break down and disperse recombination fractions in heavy oil, and decrease heavy oil viscosity. Post-activation, the utilization of residual oil in medium and small pores significantly improved, particularly in residual oil with film-like and cluster-like structures. This process slowed the decline rate of light/heavy fractions in heavy oil. The heavy oil activator effectively enhanced the sweep volume and displacement efficiency of heavy oil, thereby playing a crucial role in improving recovery rates in heavy oil reservoirs. Ma [24] synthesized a novel low-molecular-weight amphiphilic viscosity reducer. Employing Materials Studio 5.0 software, Ma explored the synthesis feasibility and viscosity reduction mechanism of this heavy oil viscosity reducer through a molecular dynamics perspective. The molecular dynamics simulations revealed that the addition of the viscosity reducer altered the potential energy, non-potential energy, density, and hydrogen bond distribution within the heavy oil. Specifically, the benzene ring in butadiene benzene became embedded in the asphaltene's interlayer structure, effectively weakening the heavy oil's network configuration. The results of numerous studies have shown that amphiphilic polymers prepared by introducing hydrophobic monomers into acrylamide-based copolymers have good effects on improving heavy oil recovery.

In this study, based on the cost-effectiveness of materials and the difficulty of synthesis, we selected acrylamide and 2-acrylamido-2-methylpropane sulfonic acid commonly used in oil fields as hydrophilic monomers and selected long-chain lauryl acrylate and benzyl acrylate with benzene ring as hydrophobic monomers to prepare amphiphilic polymers PAALB through free radical polymerization. The addition of acrylamide and 2-acrylamido-2-methylpropanesulfonic acid effectively improved the viscosity, temperature resistance and salt resistance of the polymer solution. The addition of monomers with long chains provided the polymers with interfacial activity. The addition of monomers containing benzene rings facilitates the dispersion of heavy oil components in the polymer, thus enhancing the interaction between the polymer and the heavy oil. The amphiphilic polymer has excellent interfacial activity, emulsifying ability and viscoelasticity, which can both reduce the viscosity of the oil phase and increase the viscosity of the aqueous phase. This enhancement changes the water-oil mobility ratio, increases sweep volumes, and contributes to the overall enhancement of heavy oil recovery. In addition, an important capability of our synthesized amphiphilic polymers is the ability to achieve self-emulsification of heavy oil. However, in the reports on spontaneous emulsification of crude oil, the viscosity of most crude oil is lower than 2000 mPa·s, and the spontaneous emulsification of heavy oil above 10,000 mPa·s is not reported [25–29]. Therefore, this amphiphilic polymer has better applicability to the self-emulsification and viscosity reduction of high-viscosity heavy oils.

2. Materials and Methods

2.1. Materials

The reagents used include acrylamide (AM, 99% pure, CAS No. 79-06-1), 2-acrylamido-2-methylpropansufonic acid (AMPS, purity 98%, CAS No. 15214-89-8), lauryl acrylate (LA, purity 90%, CAS No. 2156-97-0), benzyl acrylate (BA, purity 97%, CAS No. 2495-35-4), 2,2′-azobis (2-methylpropionamidine) dihydrochloride (AIBA, purity 97%, CAS No. 2997-92-4), sodium hydroxide (NaOH, purity 98%, CAS No. 1310-73-2), iron sulfate heptahydrate (FeSO$_4$·7H$_2$O, purity 99%,CAS No. 7782-63-0), hydrogen peroxide (H$_2$O$_2$, 30% aqueous solution, CAS No. 7722-84-1), edetate disodium (EDTA, purity 98%, CAS No. 6381-92-6), ethanol (purity 99.7%, CAS No. 64-17-5), and sodium dodecyl sulfate (SDS, purity 98%, CAS No. 151-21-3). These were sourced from Aladdin company (Shanghai, China).

Partially hydrolyzed polyacrylamide (HPAM) with a relative molecular weight of 1.2×10^7 and hydrolysis degree of 20% was obtained from Zhangjiakou Shengda Polymer Co., Ltd. (Zhangjiakou, China). The heavy oil used in the experiments was sourced from Chunfeng Oilfield, and its physical properties are presented in Table 1. The mineralization composition of the brine water is shown in Table 2.

Table 1. Properties and composition of Chunfeng Heavy Oil.

Relative Density	Viscosity (50 °C)	Saturates	Aromatics	Resin	Asphaltene	Wax
0.983 g/cm^3	14,315 mPa·s	46.17%	17.64%	10.11%	10.69%	2.62%

Table 2. Composition of Brine water (mg/L).

Na$^+$	K$^+$	Ca^{2+}	Mg^{2+}	Cl$^-$	SO$_4{}^{2-}$	HCO$_3{}^-$	Total Dissolved Solids
11,500	126	6600	62	29,200	260	652	48,400

2.2. Synthesis of PAALB

The synthesis process of the PAALB polymer is illustrated in Figure 1. PAALB, a quaternary copolymer, was synthesized through free radical copolymerization. The process involved using AM, AMPS, LA, and BA as the raw materials. The monomers were measured and placed in a 250 mL three-necked flask equipped with a mechanical overhead stirrer and subjected to N$_2$ purge gas. The molar ratio of AM:AMPS:LA:BA was 6:2:1:1. Deionized water was added to the flask to maintain the total monomer concentration in the solution at 30%. SDS (at 0.3% of the total monomer) was introduced to facilitate the dissolution of the hydrophobic monomer, and the solution's pH was adjusted to 6 using an aqueous NaOH solution. Following a 30-min N$_2$ purge, the solution was cooled to a temperature within 0–10 °C. Subsequently, 50 mg of AIBA, along with 300 ppm of EDTA, 500 ppm of FeSO$_4$, and 400 ppm of H$_2$O$_2$, was added as an initiator, and the mixture was stirred for 30 min. This process resulted in a milky white viscous liquid. The temperature was then increased to 60 °C and maintained for 4 h, yielding the final product, PAALB. The product was washed with ethanol to remove impurities. It was subsequently dried in a vacuum oven at 60 °C for 24 h to obtain a solid product, which was then ground into a powder and stored [18]. The final polymer yield was 88.67%.

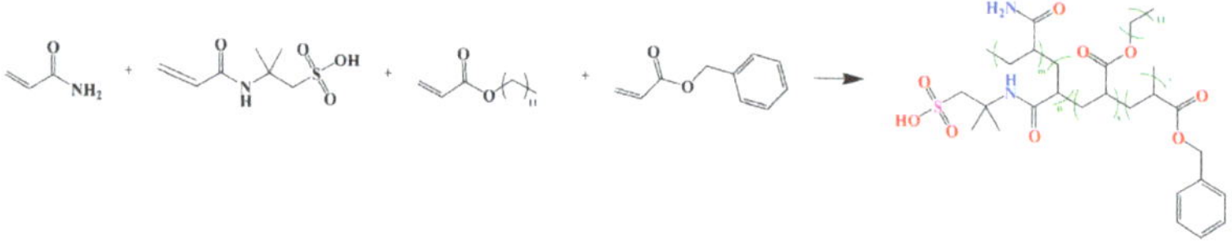

Figure 1. Synthesis of the amphiphilic polymer PAALB.

2.3. Polymer Characterization

The polymer powder was blended with KBr and compressed into a thin film. Subsequently, the Fourier transform infrared spectrum (FT-IR) was obtained using the Thermo Fischer Scientific Nicolet IS5 spectrometer (Thermo Fisher Scientific, Waltham, MA, USA).

A polymer solution with a concentration of 20 mg/mL was prepared by dissolving the polymer in D_2O. The solution was added to the nuclear magnetic tube for scanning. The ^1H nuclear magnetic resonance (NMR) spectroscopic analysis used the Bruker AVANCE III 600 NMR spectrometer (Bruker Corporation, New York, NY, USA).

C, H, O, N, and S quantities were determined using the Vario EL cube and Vario MACRO cube elemental analyzer (Elementar Germany, Frankfurt, Germany).

Gel permeation chromatography (GPC) was performed using an Agilent 1260 Infinity II LC liquid chromatograph equipped with an Agilent RID G1362A detector (Agilent Technologies Inc, Santa Clara, CA, USA) on a Waters Ultrahydrogel 300 × 7.8 mm 500-250-120A column. In this study, 0.1 mol/L $NaNO_3$ aqueous solution was used as the mobile phase at a flow rate of 1 mL/min. The testing temperature was 40 °C, and the polymer sample concentration was 5000 mg/L.

2.4. Apparent Viscosity Measurement

The viscosity of the polymer solution to crude oil was determined using a HAAKE MARS III rheometer (Thermo Fisher Scientific, Waltham, MA, USA). The CC24Ti tumbler measuring system was selected for viscosity determination in a CR test mode during viscosity tests. The sample was sheared at $7.34 \, s^{-1}$ for 300 s at 50 °C, and the measurements were taken every 30 s. The viscosity–temperature curve was determined in a CR test mode at a shear rate of $7.34 \, s^{-1}$, with recordings made from 20 to 100 °C at a step of 3 °C per 60 s for a total test time of 1600 s.

2.5. Critical Association Concentration Measurement

This research employed pyrene as a fluorescence spectroscopy probe to investigate the hydrophobic association behavior of amphiphilic polymers. In the experiment, a pyrene ethanol solution of 2.5×10^{-3} mol/L was added to various polymer solution concentrations to achieve a final pyrene concentration of 5×10^{-5} mol/L (a concentration that had minimal impact on solution properties). The solutions were sealed, stirred, and left for 24 h. The fluorescence spectrum of pyrene was recorded using a Hitachi F4600 fluorescence spectrophotometer (HITACHI, Tokyo, Japan), with an excitation wavelength of 335 nm and an emission spectrum recording range of 350–450 nm. The I_1/I_3 ratio reflected the change in polarity within the microenvironment surrounding pyrene, enabling the determination of hydrophobic association strength in the amphiphilic polymer solutions.

2.6. Polymer Self-Emulsification Ability Evaluation

For emulsification analysis, a 30 mL glass bottle was used, combining 4.5 mL of 5000 mg/L polymer solution and 10.5 mL of heavy oil, resulting in an oil–water ratio of 7:3. The bottle was inverted 50 times, and the color of the solution was observed. A uniform black solution after inversion indicated successful emulsification of heavy oil and the polymer [25–27].

2.7. Viscosity Reduction Performance Measurement

Polymer solutions of varying concentrations were prepared using brine water. Under conditions of 50 °C and a 7:3 oil–water ratio, the glass bottle was inverted 50 times, and the viscosity of the emulsion post-inversion was measured to evaluate the polymer's heavy oil viscosity reduction performance. The viscosity reduction rate was calculated using the following formula:

$$\Phi = \frac{\mu_0 - \mu}{\mu_0} \times 100\% \tag{1}$$

where Φ is the viscosity reduction rate, μ_0 is the initial viscosity of heavy oil, and μ is the viscosity of oil–water emulsion after adding polymer.

2.8. Emulsion Structure Testing

For the examination of the emulsion formed by heavy oil and the polymer solution after emulsification, a small amount was drawn using a pipette and placed onto a microscope slide. A cover slip was carefully placed over the droplet using tweezers. The sample was then positioned on the stage of a Zeiss Primo Star microscope for emulsion image observation with a magnification of 100×.

2.9. Interfacial Tension Measurement

Polymer solutions with various concentrations were prepared using brine water. The interfacial tension (IFT) between the polymer solution and heavy oil was measured using the SVT20N video spinning drop tensiometer. The test was conducted at a temperature of 50 °C and a rotation speed of 6000 r/min.

2.10. Contact Angle Measurement

Core slices with neutral wetting surfaces were soaked in a 5000 mg/L polymer solution for 24 h. Then, the slices were dried in a hot air sterilizing drying oven at 80 °C. The contact angle of water droplets on the core surface was measured using an SDC200 contact angle meter (Ningbo Precious Instruments Technology Co., Ltd., Ningbo, China) to determine any changes in wettability on the core surface.

2.11. Polymer Flooding Performance
2.11.1. Sand-Packed Tube Flooding Experiments

A sand-packed tube flooding device was utilized to evaluate the oil displacement effects of HPAM and PAALB (Figure 2). The parameters of the sand-packed tube are shown in Table 3. The sand-packed tube was filled with quartz sand, saturated with water to measure the porosity, and injected with water to measure the permeability. After that, the sand-packed tube was saturated with oil and allowed to age for 24 h. Water was then injected to displace the oil until the water content exceeded 98%. Subsequently, polymer injection was performed to displace the oil until the water content exceeded 98%. The following step was subsequent water flooding. The experimental conditions were as follows: temperature of 50 °C, HPAM concentration of 15,000 mg/L (150 mPa·s), PAALB concentration of 10,000 mg/L (145 mPa·s), and injection speed of 0.2 mL/min.

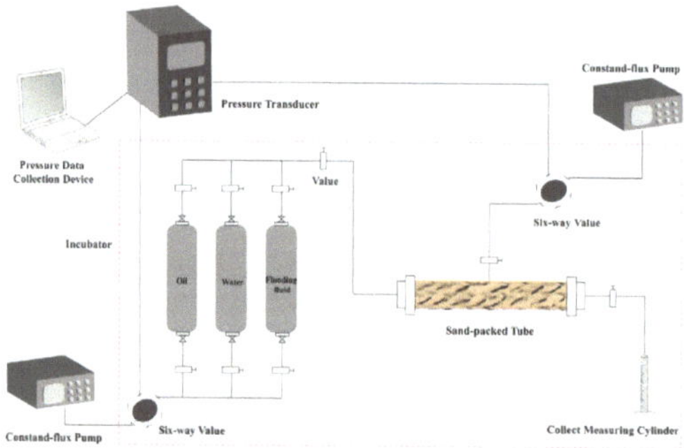

Figure 2. Schematic diagram of the sand-packed tube flooding experimental device.

Table 3. Parameters of the sand-packed tube.

Polymer	Length (cm)	Diameter (cm)	Porosity (%)	Permeability (μm²)	Oil Saturation (%)
HPAM	30.00	2.50	36.06	5.13	89.17
PAALB	30.00	2.50	35.71	5.07	92.28

2.11.2. Microscopic Model Flooding Experiments

The microscopic oil displacement effects of HPAM and PAALB were evaluated using a microscopic visualization model etched on glass (Figure 3). The experimental procedure included injecting silicone oil and aging it for 12 h to treat the pore surface of the model as an oil-wet surface. Subsequently, the micro model was saturated with oil and allowed to age for 24 h. Water was then injected to displace the oil until no oil was produced at the outlet. At this point, the polymer was injected until no oil was produced at the outlet, allowing observation of the oil displacement by the polymer. The experimental conditions were as follows: temperature of 50 °C, HPAM concentration of 15,000 mg/L (150 mPa·s), PAALB concentration of 10,000 mg/L (145 mPa·s), and injection speed of 0.12 mL/h.

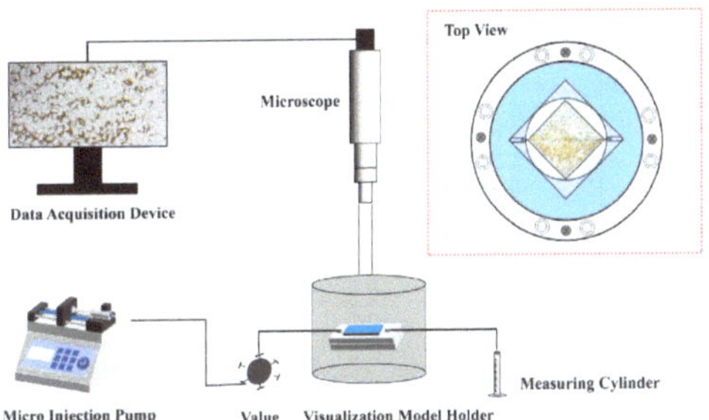

Figure 3. Schematic diagram of the microscopic visualization flooding experiment device.

3. Results and Discussion

3.1. Characterization of PAALB

3.1.1. FT-IR and ^{1}H-NMR

Figure 4a shows the infrared spectra of the monomer and polymer. The spectrum of acrylamide has distinct stretching vibrational bands of amino (–NH$_2$) at 3170 cm^{-1} and 3340 cm^{-1}. This peak slows down and shifts in the polymer spectrum, appearing at 3215 cm^{-1} and 3331 cm^{-1}. The peaks at 1075 cm^{-1} and 1232 cm^{-1} in the infrared spectra of 2-acrylamido-2-methylpropansufonic acid are attributed to the symmetric and asymmetric vibrational absorption peaks of the sulfonic acid group (–SO$_3{}^-$); this absorption peak in the polymer spectrum is displaced and appears at 1037 cm^{-1} and 1157 cm^{-1}. The peaks at 2853 cm^{-1} and 2922 cm^{-1} in the spectrum of lauryl acrylate are the stretching vibrational bands of methylene (–CH$_2$) and methyl (–CH$_3$) groups, and this peak can also be observed in the polymer spectrum. The peaks at 3067 cm^{-1} and 3081 cm^{-1} in the spectrum of benzyl acrylate are the characteristic group frequencies of the hydrocarbon bond (C–H) stretching vibration of the benzene ring, and the peaks at 696 cm^{-1} and 737 cm^{-1} are the characteristic absorption peaks of the monosubstituted benzene ring. In addition, the absorption peaks of the carbonyl group (C=O) were observed at 1773 cm^{-1} and 1671 cm^{-1}. Conclusively, the polymers' spectra contain each reacting monomer's characteristic peaks, proving that we successfully prepared the copolymer PAALB via monomer copolymerization [18,30,31].

To further confirm the molecular structure of PAALB, proton nuclear magnetic resonance spectroscopy was conducted, as displayed in Figure 4b. The observed peaks and their respective assignments are as follows: the peak at 1.46 ppm corresponds to the H of the methylene (–CH$_2$) on the long chain of lauryl acrylate. The peak at 1.49 ppm corresponds to the H of the methyl (–CH$_3$) on AMPS. The peak at 1.73 ppm corresponds to the H of the secondary methyl on the polymer main chain. The peak at 2.19 ppm corresponds to the H of the methine on the polymer main chain. The peak at 3.41 ppm corresponds to the H of the methylene connected to S on AMPS. The peak near 5.67 ppm corresponds to the H of the methine connected to the benzene ring on benzyl acrylate. The peak near 6.11 ppm corresponds to the H of the amino group (–NH$_2$) on acrylamide. The peak at 7.45 ppm corresponds to the H of the benzene ring on benzyl acrylate [32,33]. The results obtained from IR and ^{1}H-NMR confirm the successful synthesis of the intended product.

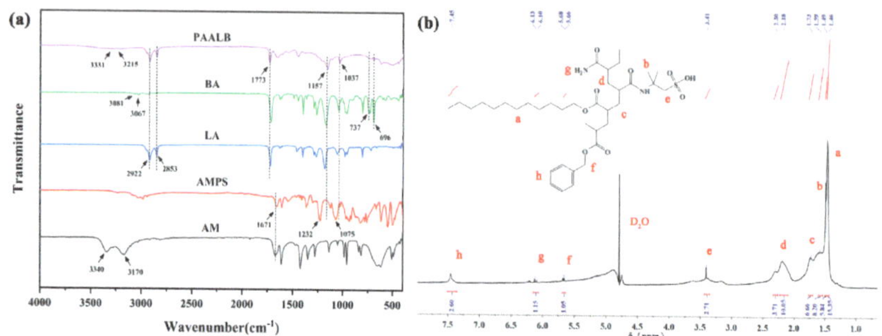

Figure 4. Structural characterization of PAALB (**a**) FT-IR; (**b**) ^{1}H-NMR.

3.1.2. Elemental Analysis

The results from the elemental analysis of the polymer are shown in Table 4. According to theoretical calculation, when the molar ratio of AM:AMPS:LA:BA is 6:2:1:1, the mass proportion of each element in the polymer is C (55.07%), H (7.57%), O (23.19%), N (9.02%) and S (5.15%), respectively. Although there is some deviation between the test results of element analysis and the theoretical calculation results, the overall results are similar, and the deviation is very slight. The test results show that the monomer composition of the polymer is consistent with the feed ratio.

Table 4. Elemental analysis results of the PAALB.

Element	C	H	O	N	S
Theoretical content (wt %)	55.07	7.57	23.19	9.02	5.15
Measurement content (wt %)	50.77	7.41	28.54	8.35	5.02

3.1.3. Relative Molecular Mass of PAALB and Its Distribution

Table 5 shows the relative molecular mass distribution of polymer PAALB. As can be seen from Table 5, the number average relative molecular mass (M_n) of the polymer is 307,883, the weight average relative molecular mass (M_w) is 679,306, the Z average relative molecular mass (M_z) is 1,048,818, the Z + 1 average relative molecular mass (M_{z+1}) is 1,333,611, and the viscosity average relative molecular mass (M_v) is 623,887. The polymer's relative molecular mass polydispersity index (PDI) was 2.21, and the test results proved the successful preparation of the polymer.

Table 5. Gel permeation chromatography results of PAALB.

M_n	M_w	M_z	M_{z+1}	M_v	PDI
307,883	679,306	1,048,818	1,333,611	623,887	2.21

3.2. Critical Association Concentration of Polymer

Figure 5 is a display of viscosity change and critical association concentration (CAC) of the polymer solution in deionized and brine water. Specifically, Figure 5a,b display the fluorescence intensity of pyrene in PAALB solutions of deionized and brine water, respectively. The emission spectrum of pyrene, excited at 335 nm, displayed five emission peaks at 373 nm, 379 nm, 384 nm, 390 nm, and 410 nm. The intensity ratio of the first vibrational peak (I_1) to the third vibrational peak (I_3) served as an indicator of the polarity of the surrounding environment for pyrene molecules. A smaller I_1/I_3 ratio suggested a higher degree of hydrophobic association. Thus, the I_1/I_3 ratio was utilized as a measure for the degree of hydrophobic association for the amphiphilic polymers. When the I_1/I_3 ratio started to decrease, it indicated the occurrence of intermolecular hydrophobic association within the polymer, and this concentration was considered the CAC for the amphiphilic polymer. Figure 5c shows the change in the I_1/I_3 ratio of the pyrene vibrational peak with polymer solution concentration, indicating the CAC of PAALB to be 3500 mg/L in deionized water solution and 5000 mg/L in brine water. Figure 5d exhibits the viscosity change of the polymer in deionized water and brine water with increasing concentration. The viscosity initially increases slowly with concentration and then rapidly rises exponentially. This behavior is attributed to the hydrophobic association between polymer molecules, which leads to a sudden increase in viscosity when the concentration exceeds the CAC. Based on curve fitting and viscosity trends, the predicted CAC concentration of PAALB in distilled water is 3400 mg/L, while in brine water, it is 4300 mg/L. These predictions closely align with the experimental results presented in Figure 5c, validating the approximate prediction of CAC using the viscosity increase method [34,35].

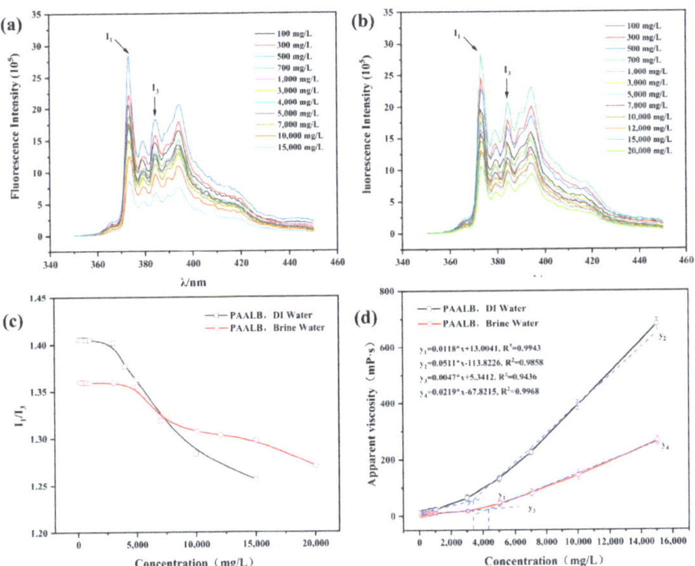

Figure 5. (**a**) Fluorescence emission spectrum of PAALB distilled water solution; (**b**) Fluorescence emission spectrum of PAALB brine water solution; (**c**) I_1/I_3 curve with PAALB Concentration; (**d**) PAALB viscosity curve with concentration.

3.3. Polymer Self-Emulsification Performance and Mechanism

3.3.1. Polymer Self-Emulsification Performance

The progression of oil–water states with increasing numbers of glass bottle inversions can be observed in Figure 6. Initially, the polymer solution and heavy oil exhibited a clear separation. After a single inversion, oil droplets appeared in the polymer solution layer. The number of the droplets in the polymer layer increased as the inversions continued, and the color of the oil–water system darkened. A uniform oil-in-water (O/W) emulsion state was observed, where the oil phase was evenly distributed as droplets within the water phase. The results of the glass bottle inversion experiment demonstrate the ability of the PAALB polymer solution to emulsify heavy oil and form an O/W emulsion under weak shear forces generated via oscillation [25,26].

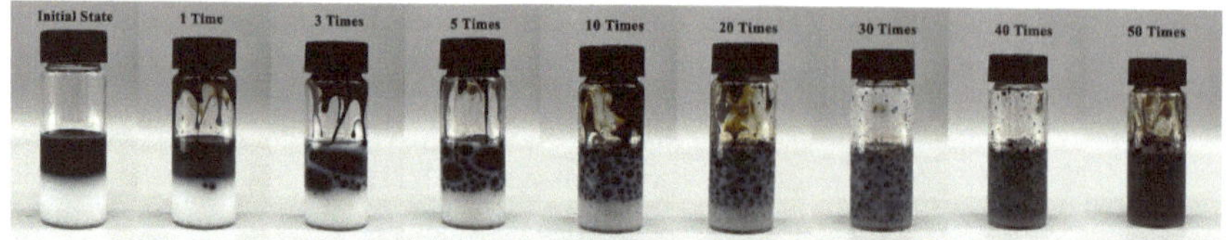

Figure 6. Oil–water state diagram after different inversion times.

3.3.2. Self-Emulsification Mechanism

The mechanisms underlying self-emulsification involve interfacial turbulence caused by Marangoni convection, the generation of negative interfacial tension values, and the diffusion and retention of chemical instability [36,37]. This paper considers the change in interfacial tension (IFT) during the dynamic self-emulsification process as the key factor influencing oil–water emulsification. Figure 7a depicts a container filled with deionized water, where water droplets in the liquid phase and at the gas-liquid interface experience different resultant forces. While water droplets in the liquid phase maintain force balance, those at the interface are subjected to a resultant force that directs them inward, causing the droplets to tend to escape the interface and enter the bulk phase. Similarly, individual droplets extracted from the liquid phase exhibit different forces acting on the surface and interior molecules. The resultant force on the surface molecules directs inward, maintaining the droplet's spherical shape. Cutting a large droplet in half yields two smaller droplets, as the interfacial tension along the cutting line contracts the surface of the cut droplet, sustaining its spherical shape. This natural phenomenon is the result of interfacial tensions. Consequently, the interfacial tension at the oil–water interface promotes the contraction of water and oil droplets. High interfacial tension increases the tendency for droplet contraction. Introducing surfactants reduces the oil–water interfacial tension, diminishing the contraction tendency. As illustrated in Figure 7b, it can be noticed that as the interfacial tension is significantly increased, the oil and water remain immiscible; thus, an enormous amount of external energy is needed to overcome this interfacial tension for the mixing to occur. When the interfacial tension is significantly reduced, minimal shearing or disturbances to the system can overcome this interfacial tension and promote oil–water mixing. When the interfacial tension becomes negative, the force driving the contraction of the oil–water interface transforms into a force facilitating the diffusion of the two phases, thus their spontaneous mixing. Figure 7c depicts the process of spontaneous emulsification of the heavy oil and polymer solution when the glass bottle is inverted. This phenomenon occurs due to the polymer solution's ability to reduce the interfacial tension between oil and water [38].

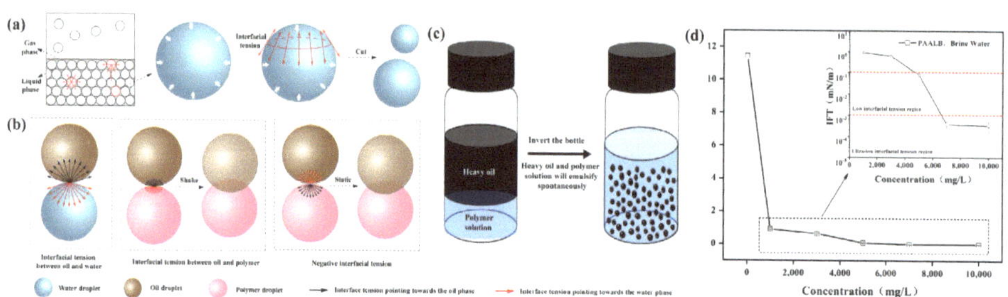

Figure 7. (**a**,**b**) Schematic diagram of the dynamic spontaneous emulsification mechanism; (**c**) self-emulsification of heavy oil with polymer induced by inverted bottle; (**d**) curve of interfacial tension with polymer concentration.

As shown in Figure 7d, oil–water interfacial tension decreases significantly with increasing polymer solution concentration. The oil–water interfacial tension without polymer addition was 11.43 mN/m. Upon reaching a PAALB concentration of 3000 mg/L, the oil–water interfacial tension decreased to 0.61 mN/m, and at a concentration of 5000 mg/L, it reduced further to 0.07 mN/m, falling within the low interfacial tension region. A solution concentration of 10,000 mg/L produced an oil–water interfacial tension of 3.32×10^{-4} mN/m, signifying an ultra-low interfacial tension. The significant reduction in interfacial tension enables oil–water self-emulsification with minimal disturbance. The experimental results of self-emulsification confirm that achieving oil–water spontaneous emulsification is possible by reducing interfacial tension [16,31].

3.4. Heavy Oil Viscosity Reduction Performance of Polymer

Figure 8a is a display of experimental results from oil–water emulsions' viscosity and viscosity reduction rate at different polymer concentrations, considering deionized and brine water conditions. As the polymer concentration increases, the emulsion viscosity initially decreases and then increases, while the viscosity reduction rate exhibits an inverse trend. Notably, the viscosity reduction effect is more pronounced in brine water. Under brine water conditions, when the polymer concentration is 1000 mg/L, the emulsion viscosity is 7307 mPa·s, with a viscosity reduction rate of 48.96%. At this concentration, the emulsification effect is limited due to the low polymer concentration. However, as the polymer concentration rises to 3000 mg/L, the emulsion viscosity drops to 109 mPa·s, and the viscosity reduction rate reaches 99.24%. From this observation, it can be deduced that increasing the concentration enhances the emulsification effect and reduces the viscosity. At a concentration of 10,000 mg/L, the emulsion viscosity rises to 201 mPa·s with a viscosity reduction rate of 98.60%. The observed decrease and subsequent increase in emulsion viscosity and the increase and subsequent drop in the viscosity reduction rate are attributed to the increased viscosity of the water phase with higher polymer concentrations. The viscosity of O/W emulsion is determined by the water phase viscosity, where higher water phase viscosity leads to higher emulsion viscosity [39–41]. This phenomenon also explains why the polymer exhibits better viscosity reduction performance in brine water than in deionized water. Figure 8b displays a microscopic image of the oil–water emulsion, revealing the O/W emulsion structure formed by the polymer and heavy oil. This O/W emulsion structure significantly reduces the oil–water system's viscosity, effectively enhancing the heavy oil's fluidity.

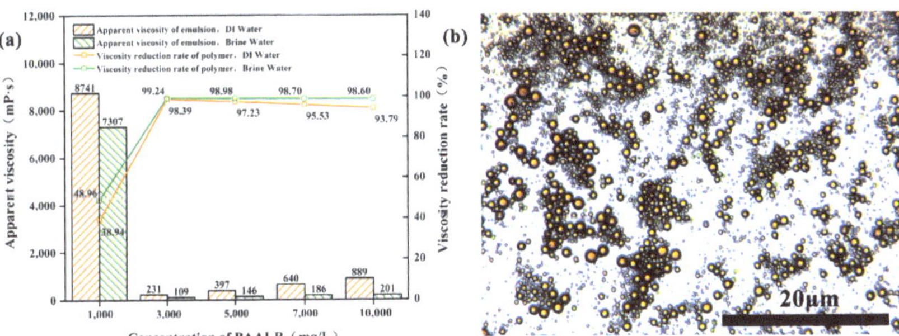

Figure 8. (**a**) Heavy oil viscosity reduction performance; (**b**) emulsion microscopic morphology.

Figure 9a,b depict the flow state of heavy oil without the addition of polymer. These images show that heavy oil exhibits poor fluidity. In stark contrast, Figure 9c,d show the flow state of heavy oil forming an emulsion when the polymer is added. The emulsion demonstrates exhibiting good fluidity. In Figure 9e, the viscosity–temperature profiles of both heavy oil and the heavy oil–polymer emulsion exhibit a declining trend as the temperature increases. However, an interesting observation arises with the emulsion. At

89 °C, there is a significant surge in viscosity, followed by a gradual decrease. This abrupt increase is attributed to the emulsion's instability induced by elevated temperature, which weakens the oil–water interfacial film. Consequently, small droplets coalesce into larger ones, resulting in increased emulsion particle size and viscosity.

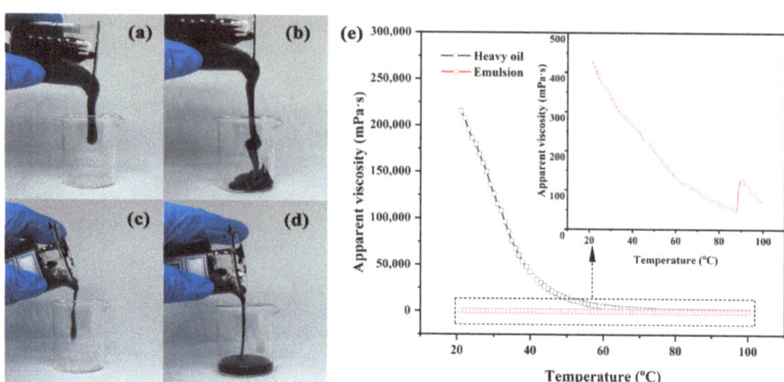

Figure 9. (**a,b**) Flowability of heavy oil; (**c,d**) flowability of emulsion; (**e**) viscosity–temperature curves of heavy oil and emulsion.

3.5. Wettability Change

Figure 10 illustrates the change in wettability of the artificial core surface before and after polymer treatment. Initially, the core surface exhibits a contact angle of 106.55°, indicating neutral wetting. However, after polymer treatment, the contact angle reduces to 34.95°, indicating a water-wet surface. The polymer treatment alters the wettability by adsorbing polymer molecules onto the core surface, with hydrophobic groups facing inward and hydrophilic groups facing outward. This transformation promotes water spreading and wetting the rock surface, facilitating the detachment of oil droplets from the core surface, and realizing oil–water replacement [19,42].

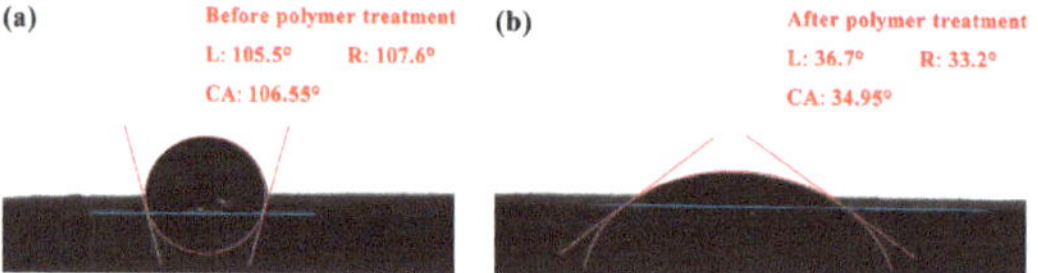

Figure 10. Contact angle of water with core surfaces. (**a**) before polymer treatment; (**b**) after polymer treatment.

3.6. Polymer Oil Displacement Performance

3.6.1. Sand-Packed Tube Flooding Experiments

Figure 11 presents the oil displacement curve of HPAM and PAALB, comprising three stages: water flooding, polymer flooding, and subsequent water flooding. In Figure 11a, during the water injection stage, the pressure difference initially rises to 2429 kPa before dropping rapidly to 220 kPa due to the formation of fluid advantage channels after water flooding oil breakthrough. After injecting 1.0 PV of water, the water cut increased sharply after the breakthrough, exceeding 98%, with a water flooding recovery rate of 23.22%. In the HPAM flooding stage, the pressure difference initially rises to 781 kPa and then declines, while the rate of water content rapidly drops to 55.78% before it gradually rises. The water cut surpasses 98% after injecting 1.5 PV of HPAM solution, leading to a 28.85% increase in the HPAM flooding stage recovery rate. The subsequent water flooding stage maintains a stable pressure difference and water cut, with minimal increase in the recovery rate. Finally, the recovery rate reaches 52.85% after injecting 3.0 PV of fluid.

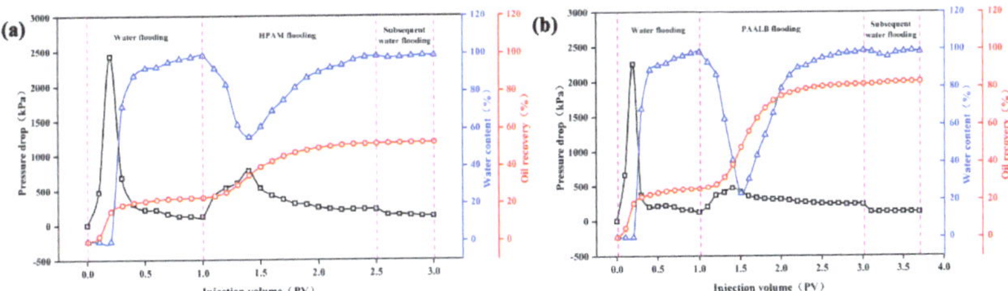

Figure 11. Polymer flooding curves; (**a**) HPAM; (**b**) PAALB.

Similarly, in Figure 11b, the water injection stage displays a rise and rapid drop in pressure difference to 2256 kPa, accompanied by a rapid increase in the water cut. After injecting 1.0 PV of water, the rate of water cut exceeds 98%, resulting in a water flooding recovery rate of 25.88%. During the PAALB flooding stage, the pressure difference initially rises to 478 kPa before declining, and the water cut drastically drops to 23.33%, followed by a gradual rise. The water cut after injection of 1.5 PV of PAALB solution was 95.44%, which increased recovery rate by 54.48%. The water cut exceeds 98% after injecting 2.0 PV of PAALB solution, leading to a 55.68% increase in the PAALB flooding recovery rate. The subsequent water flooding maintains a stable pressure difference and water cut, with a slight increase in the recovery rate. Finally, the recovery rate reaches 83.09% after injecting 3.7 PV of fluid.

3.6.2. Microscopic Model Flooding Experiments

Figure 12 illustrates the microscopic mechanism of oil displacement via HPAM and PAALB. Figure 12a,d display photos and microscopic images of the model saturated with heavy oil, revealing crude oil filling the model pores. Following oil displacement via HPAM, as shown in Figure 12b,e, more residual oil remains, and fingering phenomena occur due to the significant difference in oil–water mobility. This displacement process represents the water phase pushing the oil phase forward. Figure 12c,f depict images after oil displacement by PAALB, showcasing less residual oil in the model slice and a substantial presence of fine oil droplets. This behavior suggests that PAALB disperses large oil droplets into tiny oil droplets during the displacement process, facilitating subsequent fluid displacement [43–45].

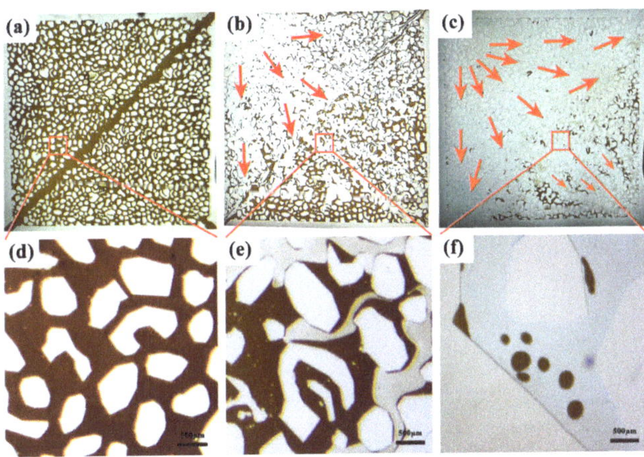

Figure 12. (**a–c**) Model image of oil saturated, HPAM flooding, and PAALB flooding; (**d–f**) microscopic image of oil saturated, HPAM flooding, and PAALB flooding (Scale Bars, 500 μm).

3.7. Amphiphilic Polymer Flooding Mechanism

In Figure 13, the schematic diagram showcases the mechanism of oil replacement, self-emulsification, and oil displacement by the amphiphilic polymer PAALB. Figure 13a depicts the distribution of heavy oil on the reservoir rock surface and within the rock pores. Figure 13b illustrates the oil replacement process during the initial stage of polymer injection. The polymer first adsorbs onto the rock surface, alters the wettability, and detaches oil droplets from the rock surface, dispersing the oil phase in the polymer solution. Figure 13c illustrates the process of self-emulsification during polymer oil displacement. Shear action during reservoir migration disturbs the heavy oil and polymer, leading to the self-emulsification of oil and water. The polymer molecules disperse and encapsulate large oil blocks, forming numerous tiny oil droplets and generating O/W type emulsion. Figure 13d illustrates the state after the polymer solution interacts with heavy oil. The polymer peels off and emulsifies the heavy oil from the rock surface, transforming it from initial large oil droplets to tiny oil droplets. This phenomenon improves fluidity and enhances the displacement and production of the oil phase, ultimately increasing the heavy oil recovery.

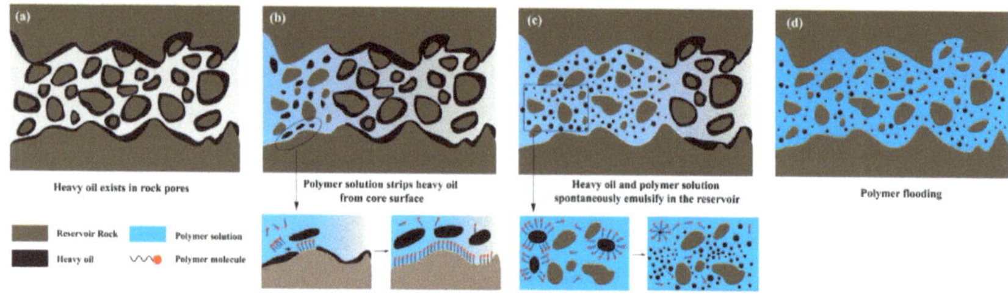

Figure 13. Schematic diagram of polymer mechanism for heavy oil recovery.

4. Conclusions

In summary, we synthesized an amphiphilic polymer, PAALB, through free radical polymerization, with the aim of enhancing heavy oil recovery. This versatile polymer offers multiple functionalities, including the reduction of heavy oil viscosity, oil–water displacement, and polymer flooding. A solution of 5000 mg/L PAALB effectively reduces the contact angle of water droplets on the core surface from 106.55° to 34.95°. This transition from oil-wet to water-wet conditions enables oil–water displacement. Moreover, under a 7:3 oil–water ratio, a 10,000 mg/L PAALB solution, using the glass bottle inversion method, decreases the viscosity of heavy oil from 14,315 mPa·s to 201 mPa·s, achieving an impressive viscosity reduction rate of 98.60%. In sand-packed tube flooding experiments with the same viscosity and injection volume of 1.5 PV, PAALB enhances the recovery rate by 25.63% compared to the traditional HPAM polymer.

High interfacial tension induces the contraction of oil droplets and water droplets at the oil–water interface, resulting in their separation. Notably, a 10,000 mg/L PAALB solution reduces the oil–water interfacial tension to an ultra-low level of 3.32×10^{-4} mN/m. This outcome mitigates the contraction tendency of droplets at the interface, promoting oil–water mixing and emulsification. It enables the self-emulsification of heavy oil and the polymer solution under weak shear conditions, transforming high-viscosity heavy oil into low-viscosity oil-in-water (O/W) emulsion, thereby improving heavy oil fluidity. Furthermore, PAALB can increase the viscosity of the water phase through hydrophobic association. This enhancement ultimately improves the flow ratio, sweep volume, and, most importantly, heavy oil recovery.

Author Contributions: D.F.: Methodology, Investigation, Writing—original draft. J.G.: Conceptualization, Funding acquisition, Resources. R.X.: Methodology, Writing—review and editing. X.Z.: Validation. C.K.: Visualization. W.K.: Writing—review and editing. All authors have read and agreed to the published version of the manuscript.

Funding: This study received support and funding from the National Natural Science Foundation of China (No. 52174047), Sinopec Project (No. P23138), and Sinopec Project (No. P19018-2).

Institutional Review Board Statement: The study did not require ethical approval.

Data Availability Statement: The data used to support the findings of this study are available from the corresponding author upon request.

Conflicts of Interest: The author declare that the research was conducted in the absence of any commercial or financial relationships that could be construed as potential conflict of interest.

References

1. Dong, X.; Liu, H.; Chen, Z.; Wu, K.; Lu, N.; Zhang, Q. Enhanced oil recovery techniques for heavy oil and oilsands reservoirs after steam injection. *Appl. Energy* **2019**, *239*, 1190–1211. [CrossRef]
2. Wu, Z.; Huiqing, L.; Cao, P.; Yang, R. Experimental investigation on improved vertical sweep efficiency by steam-air injection for heavy oil reservoirs. *Fuel* **2021**, *285*, 119138. [CrossRef]
3. Shah, A.; Fishwick, R.; Wood, J.; Leeke, G.; Rigby, S.; Greaves, M. A review of novel techniques for heavy oil and bitumen extraction and upgrading. *Energy Environ. Sci.* **2010**, *3*, 700–714. [CrossRef]
4. Guo, K.; Li, H.; Yu, Z. In-situ heavy and extra-heavy oil recovery: A review. *Fuel* **2016**, *185*, 886–902. [CrossRef]
5. Sun, J.H.; Zhang, F.S.; Wu, Y.W.; Liu, G.L.; Li, X.N.; Su, H.M.; Zhu, Z.Y. Overview of emulsified viscosity reducer for enhancing heavy oil recovery. *IOP Conf. Ser. Mater. Sci. Eng.* **2019**, *479*, 012009. [CrossRef]
6. Wang, X.; Wang, F.; Taleb, M.A.M.; Wen, Z.; Chen, X. A Review of the Seepage Mechanisms of Heavy Oil Emulsions during Chemical Flooding. *Energies* **2022**, *15*, 8397. [CrossRef]
7. Kumar Sharma, V.; Singh, A.; Tiwari, P. An experimental study of pore-scale flow dynamics and heavy oil recovery using low saline water and chemical flooding. *Fuel* **2023**, *334*, 126756. [CrossRef]
8. Cao, H.; Li, Y.; Gao, W.; Cao, J.; Sun, B.; Zhang, J. Experimental investigation on the effect of interfacial properties of chemical flooding for enhanced heavy oil recovery. *Colloids Surf. A Physicochem. Eng. Asp.* **2023**, *677*, 132335. [CrossRef]
9. Wang, H.; Wei, B.; Hou, J.; Liu, Y.; Du, Q. Heavy oil recovery in blind-ends during chemical flooding: Pore scale study using microfluidic experiments. *J. Mol. Liq.* **2022**, *368*, 120724. [CrossRef]
10. Liu, J.; Liu, S.; Zhong, L.; Yuan, S.; Wang, Q.; Wei, C. Study on the emulsification characteristics of heavy oil during chemical flooding. *Phys. Fluids* **2023**, *35*, 53330. [CrossRef]
11. Liu, Z.; Cheng, H.; Li, Y.; Li, Y.; Chen, X.; Zhuang, Y. Experimental Investigation of Synergy of Components in Surfactant/Polymer Flooding Using Three-Dimensional Core Model. *Transp. Porous Media* **2018**, *126*, 317–335. [CrossRef]
12. Li, W.; Dai, C.; Ouyang, J.; Aziz, H.; Tao, J.; He, X.; Zhao, G. Adsorption and retention behaviors of heterogeneous combination flooding system composed of dispersed particle gel and surfactant. *Colloids Surf. A Physicochem. Eng. Asp.* **2018**, *538*, 250–261. [CrossRef]
13. Wang, Z.; Yu, T.; Lin, X.; Wang, X.; Su, L. Chemicals loss and the effect on formation damage in reservoirs with ASP flooding enhanced oil recovery. *J. Nat. Gas Sci. Eng.* **2016**, *33*, 1381–1389. [CrossRef]
14. Wang, J.; Yuan, S.; Shi, F.; Jia, X. Experimental study of chemical concentration variation of ASP flooding. *Sci. China Ser. E Technol. Sci.* **2009**, *52*, 1882–1890. [CrossRef]
15. Xiao, H.M.; Sun, L.H.; Kou, H.H. Mass Transfer Mechanisms of ASP Flooding in Porous Media. *Adv. Mater. Res.* **2012**, *550–553*, 2738–2744. [CrossRef]
16. Li, J.; Wang, Q.; Liu, Y.; Wang, M.; Tan, Y. Long Branched Chain Amphiphilic Copolymers: Synthesis, Properties, and Application in Heavy Oil Recovery. *Energy Fuels* **2018**, *32*, 7002–7010. [CrossRef]
17. Qin, X.; Zhu, S.; Shi, Q.; Li, C.; Baykara, H. Synthesis and Properties of a Dendrimer Amphiphilic Polymer as Enhanced Oil Recovery Chemical. *J. Chem.* **2023**, *2023*, 4271446. [CrossRef]
18. Chen, M.; Chen, W.; Wang, Y.; Ding, M.; Zhang, Z.; Liu, D.; Mao, D. Hydrogen-Bonded amphiphilic polymer viscosity reducer for enhancing heavy oil Recovery: Synthesis, characterization and mechanism. *Eur. Polym. J.* **2022**, *180*, 111589. [CrossRef]
19. Liu, Z.; Mendiratta, S.; Chen, X.; Zhang, J.; Li, Y. Amphiphilic-Polymer-Assisted Hot Water Flooding toward Viscous Oil Mobilization. *Ind. Eng. Chem. Res.* **2019**, *58*, 16552–16564. [CrossRef]
20. Yang, H.; Zhang, H.; Zheng, W.; Zhou, B.; Zhao, H.; Li, X.; Zhang, L.; Zhu, Z.; Kang, W.; Ketova, Y.A.; et al. Effect of hydrophobic group content on the properties of betaine-type binary amphiphilic polymer. *J. Mol. Liq.* **2020**, *311*, 113358. [CrossRef]
21. Tian, S.; Gao, W.; Liu, Y.; Kang, W. Study on the stability of heavy crude oil-in-water emulsions stabilized by two different hydrophobic amphiphilic polymers. *Colloids Surf. A Physicochem. Eng. Asp.* **2019**, *572*, 299–306. [CrossRef]
22. Zhu, Z.; Kang, W.; Sarsenbekuly, B.; Yang, H.; Dai, C.; Yang, R.; Fan, H. Preparation and solution performance for the amphiphilic polymers with different hydrophobic groups. *J. Appl. Polym. Sci.* **2017**, *134*. [CrossRef]

23. Zhi, J.; Liu, Y.; Chen, J.; Jiang, N.; Xu, D.; Bo, L.; Qu, G. Performance Evaluation and Oil Displacement Effect of Amphiphilic Polymer Heavy Oil Activator. *Molecules* **2023**, *28*, 5257. [CrossRef] [PubMed]
24. Ma, C.; Liu, X.; Xie, L.; Chen, Y.; Ren, W.; Gu, W.; Zhang, M.; Zhou, H. Synthesis and Molecular Dynamics Simulation of Amphiphilic Low Molecular Weight Polymer Viscosity Reducer for Heavy Oil Cold Recovery. *Energies* **2021**, *14*, 6856. [CrossRef]
25. Li, Z.; Wu, H.; Yang, M.; Jiang, J.; Xu, D.; Feng, H.; Lu, Y.; Kang, W.; Bai, B.; Hou, J. Spontaneous Emulsification via Once Bottom-Up Cycle for the Crude Oil in Low-Permeability Reservoirs. *Energy Fuels* **2018**, *32*, 3119–3126. [CrossRef]
26. Li, Z.; Kang, W.; Bai, B.; Wu, H.; Gou, C.; Yuan, Y.; Xu, D.; Lu, Y.; Hou, J. Fabrication and Mechanism Study of the Fast Spontaneous Emulsification of Crude Oil with Anionic/Cationic Surfactants as an Enhanced Oil Recovery (EOR) Method for Low-Permeability Reservoirs. *Energy Fuels* **2019**, *33*, 8279–8288. [CrossRef]
27. Liu, Q.; Dong, M.; Yue, X.; Hou, J. Synergy of alkali and surfactant in emulsification of heavy oil in brine. *Colloids Surf. A Physicochem. Eng. Asp.* **2006**, *273*, 219–228. [CrossRef]
28. Wu, T.; Firoozabadi, A. Surfactant-Enhanced Spontaneous Emulsification Near the Crude Oil–Water Interface. *Langmuir* **2021**, *37*, 4736–4743. [CrossRef]
29. Li, Z.; Lu, Y.; Wu, H.; Yang, M.; Feng, H.; Xu, D.; Hou, J.; Kang, W.; Yang, H.; Zhao, Y.; et al. A Novel Ultra-Low IFT Spontaneous Emulsification System for Enhanced Oil Recovery in Low Permeability Reservoirs. In Proceedings of the SPE EOR Conference at Oil and Gas West Asia, Muscat, Oman, 26–28 March 2018.
30. Wang, Y.; Li, M.; Hou, J.; Zhang, L.; Jiang, C. Design, synthesis and properties evaluation of emulsified viscosity reducers with temperature tolerance and salt resistance for heavy oil. *J. Mol. Liq.* **2022**, *356*, 118977. [CrossRef]
31. Li, Q.; Wang, X.-D.; Li, Q.-Y.; Yang, J.-J.; Zhang, Z.-J. New Amphiphilic Polymer with Emulsifying Capability for Extra Heavy Crude Oil. *Ind. Eng. Chem. Res.* **2018**, *57*, 17013–17023. [CrossRef]
32. Wu, R.; Yan, Y.; Li, X.; Tan, Y. Preparation and evaluation of double-hydrophilic diblock copolymer as viscosity reducers for heavy oil. *J. Appl. Polym. Sci.* **2022**, *140*, e53278. [CrossRef]
33. Zheng, C.; Fu, H.; Wang, Y.; Zhang, T.; Duan, W. Preparation and mechanism of hyperbranched heavy oil viscosity reducer. *J. Pet. Sci. Eng.* **2021**, *196*, 107941. [CrossRef]
34. Mao, J.; Cao, H.; Zhang, H.; Du, A.; Xue, J.; Lin, C.; Yang, X.; Wang, Q.; Mao, J.; Chen, A. Design of salt-responsive low-viscosity and high-elasticity hydrophobic association polymers and study of association structure changes under high-salt conditions. *Colloids Surf. A Physicochem. Eng. Asp.* **2022**, *650*, 129512. [CrossRef]
35. Pu, W.; Jiang, F.; Wei, B.; Tang, Y.; He, Y. Influences of structure and multi-intermolecular forces on rheological and oil displacement properties of polymer solutions in the presence of Ca^{2+}/Mg^{2+}. *RSC Adv.* **2017**, *7*, 4430–4436. [CrossRef]
36. Solans, C.; Morales, D.; Homs, M. Spontaneous emulsification. *Curr. Opin. Colloid Interface Sci.* **2016**, *22*, 88–93. [CrossRef]
37. López-Montilla, J.C.; Herrera-Morales, P.E.; Pandey, S.; Shah, D.O. Spontaneous Emulsification: Mechanisms, Physicochemical Aspects, Modeling, and Applications. *J. Dispers. Sci. Technol.* **2002**, *23*, 219–268. [CrossRef]
38. Li, Z.; Xu, D.; Yuan, Y.; Wu, H.; Hou, J.; Kang, W.; Bai, B. Advances of spontaneous emulsification and its important applications in enhanced oil recovery process. *Adv. Colloid Interface Sci.* **2020**, *277*, 102119. [CrossRef]
39. Zhang, J.; Xu, J.-Y. Apparent viscosity characteristics and prediction model of an unstable oil-in-water or water-in-oil dispersion system. *J. Dispers. Sci. Technol.* **2018**, *40*, 1645–1656. [CrossRef]
40. Luo, H.; Wen, J.; Lv, C.; Wang, Z. Modeling of viscosity of unstable crude oil–water mixture by characterization of energy consumption and crude oil physical properties. *J. Pet. Sci. Eng.* **2022**, *212*, 110222. [CrossRef]
41. Zhang, J.; Yuan, H.; Zhao, J.; Mei, N. Viscosity estimation and component identification for an oil-water emulsion with the inversion method. *Appl. Therm. Eng.* **2017**, *111*, 759–767. [CrossRef]
42. Zhang, Q.; Mao, J.; Liao, Y.; Mao, J.; Yang, X.; Lin, C.; Wang, Q.; Huang, Z.; Xu, T.; Liu, B.; et al. Salt resistance study and molecular dynamics simulation of hydrophobic-association polymer with internal salt structure. *J. Mol. Liq.* **2022**, *367*, 120520. [CrossRef]
43. Liu, J.; Zhong, L.; Zewen, Y.; Liu, Y.; Meng, X.; Zhang, W.; Zhang, H.; Yang, G.; Shaojie, W. High-efficiency emulsification anionic surfactant for enhancing heavy oil recovery. *Colloids Surf. A Physicochem. Eng. Asp.* **2022**, *642*, 128654. [CrossRef]
44. Liu, J.; Zhong, L.; Lyu, C.; Liu, Y.; Zhang, S. W/O emulsions generated by heavy oil in porous medium and its effect on re-emulsification. *Fuel* **2022**, *310*, 122344. [CrossRef]
45. Chen, X.; Li, Y.; Liu, Z.; Trivedi, J.; Tang, Y.; Sui, M. Visualized investigation of the immiscible displacement: Influencing factors, improved method, and EOR effect. *Fuel* **2023**, *331*, 125841. [CrossRef]

 *polymers*

Article

Experimental Study on Enhanced Oil Recovery Effect of Profile Control System-Assisted Steam Flooding

Long Dong [1,*], Fajun Zhao [2], Huili Zhang [1], Yongjian Liu [2], Qingyu Huang [1], Da Liu [1], Siqi Guo [1] and Fankun Meng [1]

[1] Heilongjiang Provincial Key Laboratory of Oilfield Applied Chemistry and Technology, School of Chemical Engineering, Daqing Normal University, Daqing 163712, China
[2] Key Laboratory of Entered Oil Recovery of Education Ministry, Northeast Petroleum University, Daqing 163318, China
* Correspondence: donglong@dqnu.edu.cn

Abstract: Steam flooding is an effective development method for heavy oil reservoirs, and the steam flooding assisted by the profile control system can plug the dominant channels and further improve the recovery factor. High-temperature-resistant foam as a profile control system is a hot research topic, and the key lies in the optimal design of the foam system. In this paper, lignin was modified by sulfonation to obtain a high-temperature-resistant modified lignin named CRF; the foaming agent CX-5 was confirmed to have good high-temperature foaming ability by reducing the surface tension; the formula of the profile control system (A compound system of CRF and CX-5, abbreviated as PCS) and the best application parameters were optimized by the foam resistance factor. Finally, the effect of PCS-assisted steam flooding in enhanced oil recovery was evaluated by single sand packing tube flooding, three parallel tube flooding, and large-scale sand packing model flooding experiments. The results show that CX-5 has a good high-temperature foaming performance; the foam volume can reach more than 180 mL at 300 °C, and the half-life is more than 300 s. The optimal PCS formulation is 0.3 wt% CRF as an oil displacement agent + 0.5 wt% CX-5 as a foaming agent. The optimal gas–liquid ratio range is 1:2 to 2:1, and the high pressure and permeability are more conducive to the generation and stability of the foam. Compared with steam flooding, PCS-assisted steam flooding can improve oil recovery by 9% and 7.9% at 200 °C and 270 °C, respectively. PCS can effectively improve the heterogeneity of the reservoir, and increase the oil recovery of the three-parallel tube flooding experiment by 28.7%. Finally, the displacement results of the sand-packing model with large dimensions show that PCS can also expand the swept volume of the homogeneous model, but the effect is 9.46% worse than that of the heterogeneous model.

Keywords: steam flooding; foam; profile control; enhance oil recovery; high temperature

Citation: Dong, L.; Zhao, F.; Zhang, H.; Liu, Y.; Huang, Q.; Liu, D.; Guo, S.; Meng, F. Experimental Study on Enhanced Oil Recovery Effect of Profile Control System-Assisted Steam Flooding. *Polymers* **2023**, *15*, 4524. https://doi.org/10.3390/polym15234524

Academic Editors: Japan Trivedi, Zheyu Liu, Xin Chen, Jianbin Liu and Zhe Li

Received: 10 October 2023
Revised: 15 November 2023
Accepted: 16 November 2023
Published: 24 November 2023

1. Introduction

Oil and natural gas will still be the main energy sources in the world in the next few decades. Developing unconventional oil and gas reservoirs will guarantee stable oil and gas production. Heavy oil reservoirs have a very unfavorable water–oil mobility ratio, and steam flooding is a favorable means for their efficient development [1–5]. However, with the continuous injection of steam, channeling is intensified, influenced by factors such as well spacing, reservoir heterogeneity, crude oil properties, gravity separation, reservoir pressure and temperature, relative permeability, steam quality, mobility ratio, and fluid saturation in the formation [6,7]. Control of steam channeling to expand the swept volume is an important guarantee for the heavy oil reservoir to expand the swept volume further and improve oil recovery.

Currently, a common method to solve the problem of gravity overload and the steam channeling of steam flooding is profile control agent-assisted steam flooding. The profile control system is injected into the reservoir with steam to plug the flow channels so that the steam injected into the reservoir will enter the unswept area [8–10]. The research proposed by Stanford University [11] comprehensively considered various chemical additives related to mobility control and considered applying the profile control technology in the steam flooding process an effective method [12]. The commonly used profile control systems include association polymer [13], microgel [14–17], water-assisted gas flooding [18,19], and foam [20].

Foam as a profile control system [21,22] has the mechanism of reducing steam mobility, increasing swept coefficient, and improving oil washing efficiency, which can improve the ultimate oil recovery of the reservoir. Meanwhile, the foam will not permanently block the oil layer because of the good selectivity of its plugging property. In the early 1970s, Needham [23] and others invented the steam foam flooding technology. Since then, American Shell Oil Company, CLD Group Company, Stanford University Petroleum Research Institute, and Getty Oil Company have conducted in-depth laboratory research on steam foam flooding technology, and conducted field test research in the Midway-Sunset and North Kern Front oilfields, and have achieved initial success [24]. The U.S. Department of Energy signed cost-sharing research contracts with Petio-Lewis Corporation, CLD, and SUPRI from 1979 to 1980, and entrusted Chemical Oil Recovery Company as the steam foam flooding technology manager from the laboratory to the field [25–29]. Surfactant/foam injected into the formation, together with steam, will preferentially enter the formation with steam overburden and channeling and temporarily plug the channeling channels [23,30], diverting the steam direction to expand the swept volume [31,32]. Borchard [33], Ettinger et al. [34], and Schramm et al. [35] carried out a large number of experimental studies, respectively, and concluded that the use of foam plugging in the steam flooding process is effective.

Although the above laboratory studies have shown that foaming agents can be used as plugging agents for steam flooding, the high price and poor temperature resistance of the foaming system limit its application in the oilfields [36–39]. Therefore, the development of a foam system with excellent foamability and high-temperature stability is the key to the study of foam profile control technology [40]. The injection of a suitable concentration of lignosulfonate solution into the formation can generate a gel and plug the channeling channel. Wang et al. [41] pointed out that the glue-forming liquid prepared with lignin extracted from paper-making waste liquid as the main material can form a good gel under the conditions of pH > 9 and temperature of 50~150 °C. Therefore, this paper considers the use of cheap modified lignin to improve the temperature resistance and the foam stability of the composite system.

This paper conducts research on the design of high-temperature steam flooding assisted by a profile control system, using modified lignin to improve the temperature resistance and foam performance of the system. Firstly, temperature-resistant modified lignin (CRF) from industrial wastewater was prepared via chain scission and sulfonation reactions, and then its properties were characterized by means of infrared spectroscopy, ultraviolet spectroscopy, and a TGA test. Meanwhile, the high-temperature foaming performance, foam stability, and ability to reduce the surface tension of the solution were evaluated for the foaming agent CX-5. Through the sand pack resistance factor experiment, the optimal formulation of the profile control system (PCS, including CX-5) was optimized. Finally, a single sand pack flooding efficiency experiment, a parallel flooding experiment with three sand-packing tubes, and a large-size core flooding experiment were carried out to clarify the EOR effect of PCS-assist steam flooding. The research contained in this paper can provide an experimental basis and data supporting the optimization of the on-site steam flooding profile control system and the optimal design of the injection parameters.

2. Material and Method

2.1. Materials

Agent: Pure lignin (The molecular weight is 3.2×10^5 Da) is obtained by acid precipitation and purification of low-concentration reed papermaking black liquor from Daqing Haida Paper Co., Ltd. (Daqing, China). Brine is prepared based on the composition of formation water in Daqing Oilfield. It contained 2359.8 ppm $Na^+ + K^+$, 4278.2 ppm HCO_3^-, 1018.6 ppm Cl^-, 175.2 ppm CO_3^{2-}, 110.7 ppm SO_4^{2-}, 35.4 ppm Ca^{2+} and 22.1 ppm Mg^{2+}. High temperature resistant foaming agent CX-5 (the molecular weight is 7.4×10^3 Da) is an anionic sulfonate surfactant with an effective concentration of 15%, which is provided by Daqing Petroleum Refinery (Daqing, China); the composite solution prepared by modified lignin (CRF) and CX-5 according to a certain concentration is the profile control system (PCS). Sulfuric acid, hydrochloric acid, ethanol, NaOH, formaldehyde, nonylphenol, sodium sulfite, hydrogen peroxide, etc., were purchased from Shanghai Aladdin Company (Shanghai, China), and were analytically pure.

Crude oil and brine: The deionized water was prepared by the ultra-pure water filter of Chengdu Youpu Company (Chengdu, China). According to the formation water formula, inorganic salt was added to prepare simulated formation water, and its salinity was 8000 mg/L. The crude oil is degassed and dehydrated crude oil from Liaohe Oilfield (Panjin, China), with a viscosity of 63,670 cP at room temperature, and reduced to 10 cP when heated to 200 °C.

Sand packing model: The model is 60 cm long, 2.5 cm in diameter, and filled with 50–220 mesh quartz sand. The permeability can be adjusted by designing the proportion of quartz sand with different mesh numbers, and it can be controlled between 1.0 D–15 D.

Rectangular sand packing model: The model was welded with stainless steel, and its dimensions are as follows: length $\times$ width $\times$ height = 1200 mm $\times$ 40 mm $\times$ 120 mm. There is a heating and insulating layer outside the core model; the initial temperature of the core can be set, and experiments can be carried out under thermal insulation conditions. The quartz sand used has 60 mesh, 70~140 mesh, and 270 mesh. The model is divided into the homogeneous model (place the sand packing model vertically, compact it evenly when filling, and try to ensure that the permeability is the same everywhere); and the heterogeneous model (place the sand packing model vertically, place a layer of the screen in the middle of the model, and fill the two sides of the screen with quartz sand of different meshes to form different permeability layers). After filling, the model is placed horizontally, the upper layer has a high permeability and the lower layer has a low permeability, so as to simulate the heterogeneity of the formation.

2.2. Lignin Modification Method

(1) Oxidative chain scission of lignin. A certain amount of lignin was dissolved in NaOH solution with pH = 10 and then reacted with O_2 and 20% H_2O_2 in a 0.3 L autoclave. The reaction pressure was controlled at 0.4 Mpa, the reaction temperature was 70 °C, and the reaction time was 1.5 h. After the reaction, the product was concentrated and dried for analysis. (2) Synthesis of modified lignin surfactant (CRF). A total of 6 g of sodium sulfite and 9 g of formaldehyde were added to the above-purified sample, and 3 g of nonylphenol was weighed and dissolved in an organic solvent, then added to the autoclave, and then heated to 70 °C. After the reaction was completed and cooled to room temperature, the reactant was filtered and the precipitate was collected, vacuum-dried at 50 °C, and finally, the modified lignin solid brown powder was obtained.

2.3. Modified Lignin Properties

The sample was pressed into a film, and its infrared spectrum (wave number 4000–450 cm^{-1}) was measured with a Fourier transform infrared spectrometer (Spectrum 400, Bruker, Billerica, MA, USA, potassium bromide tablet); ethanol was used as a solvent, and an ultraviolet-visible absorption spectrometer (UV-540, Shanghai Metash Instruments Co, Ltd, Shanghai, China) was used to determine the UV spectrum of the sample (wavelength 190–380 nm); the content of carbon, hydrogen, and oxygen elements in the lignin was analyzed with an elemental analyzer (EA-300 series, Keyence, Shanghai, China); a thermogravimetric analyzer (TA Q50, Dupont, Wilmington, DE, USA) was used to perform a thermogravimetric test on the CRF (6.523 mg of the sample was weighed, with high purity N_2 used as the carrier gas. The heating rate of the thermogravimeter was 15 °C/min, and the temperature was heated from room temperature to 700 °C).

2.4. Foaming Performance of the Foaming Agent

2.4.1. Determination of Foam Stability by Bubbling Method

Determination of foam stability at room temperature and pressure: Put 20.0 mL of foaming agent solution into the container of the high-temperature and high-pressure foaming device (300 °C, 50 MPa, Huabao, Yangzhou, China), and blow 140.0 mL of N_2 at a certain speed. The change in foam height reflects the dynamic balance change in foam formation and collapse, so it is a comprehensive reflection of foam stability and foaming ability. Measure the change in the foam volume with time after the bubbling is stopped, draw the foam decay curve, and obtain the foam half-life t_0. In the experiment, a circulating water bath was used to control the temperature of the system.

Determination of high-temperature and high-pressure foam stability: Put 20.0 mL of foaming agent solution into the container of the high-temperature and high-pressure foaming device, adjust the confining pressure of the system with N_2, start the incubator to heat up to the predetermined temperature, and inject 140.0 mL at a certain speed after 1 h of N_2 at a constant-temperature (condition of temperature and pressure in the container), measure the change in the foam volume with time after the bubbling is stopped, draw the foam decay curve, and obtain the foam half-life t_1.

2.4.2. Determination of Surface Tension of Foaming Agents

CX-5 solutions with concentrations of 0.05 wt%, 0.1 wt%, 0.3 wt%, and 0.5 wt% were prepared, and the surface tension was measured by the hanging ring method (CSI-356, Sincerity, Kansas, OK, USA) at 20 °C, and the surface tension was measured by the pendant drop method (Q2000, Hangzhou, China) at high temperature and high pressure, at 200 °C and 300 °C. At the same time, the air was injected into the solution to be tested through a syringe, and the change in the surface tension value within 1000 s was recorded to obtain the dynamic surface tension change curve of CX-5.

2.5. Resistance Factor of Profile Control System

After it is clear that the foaming agent can generate stable foam, it is necessary to evaluate the actual plugging ability of the foam in the core to meet the plugging ability of the steam channeling layer. The plugging performance of the foam is usually evaluated by the resistance factor, calculated during the core flooding experiment. The foaming agent solution and air are injected simultaneously by the ISCO pump (ISCO D, 0.00001–408 mL/min, Teledyne ISCO, Lincoln, NE, USA), and the two are mixed at the inlet of the sand packing pipe. The pressure difference between the inlet and outlet of the core is measured by the pressure sensor, and the resistance factor is calculated. The specific experimental process is as follows:

(1) Design and make sand packing models of artificial quartz sand with different particle sizes to meet the requirements of the specific experimental plan for porosity and permeability. (2) After the sand pack is evacuated, the simulated formation water is saturated, and the porosity is measured. (3) Connect the experimental process according to the experimental flow chart as shown in Figure 1, and test the leakage at a pressure of 10.0 Mpa. (4) The constant temperature is 24 h, and the core flooding experiment is ready. (5) To determine the basic pressure difference P_0, set the back pressure slightly lower than the saturated vapor pressure at this temperature, turn on the feed water pump to inject water, and turn on the air feed pump (set the displacement of the feed water pump and the air feed pump according to a certain gas–liquid ratio). Then, open the bypass valve

to allow the steam and air mixture to pass through the bypass, and then close the bypass valve. When the pressure at the inlet of the sand packing pipe is slightly lower than the saturated vapor pressure at the experimental temperature, open the core inlet valve, and then open the core outlet valve to carry out the flow experiment. When the differential pressure reaches a stable level, record the basic pressure difference at both ends of the core model at this time. (6) To determine the working pressure difference P_1, replace the water with the displacement agent solution, and carry out the experiment according to the same conditions and operating steps as (5). When the pressure difference across the model reaches a steady state, record the working pressure difference across the core model. (7) Calculate the foam resistance factor, that is, the ratio of the working pressure difference to the base pressure difference. (8) Repeat steps (1)–(7) by changing the experimental parameters and conditions. The specific experimental scheme is shown in Table 1.

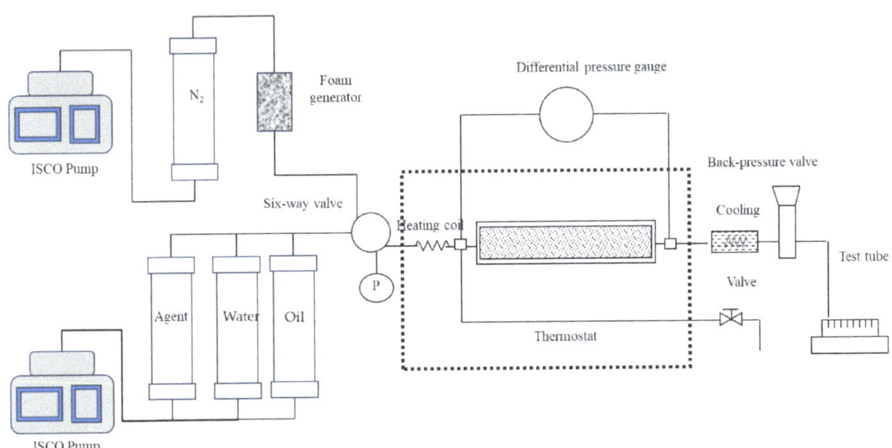

Figure 1. Schematic diagram of simulated experiment process in nitrogen foam steam flooding.

Table 1. Experimental scheme of foam resistance factor test.

Number	CRF Concentration, wt%	CX-5 Concentration, wt%	Gas and Liquid Ratio	Temperature, °C	Pressure Difference, MPa	Permeability, D	Group
1	0.2–0.5	0.3	1:1	270	5.3	About 1.0	4
2	0.3	0.2–0.6	1:1	270	5.3	About 1.0	5
3	0.3	0.5	0:1–2:1	270	5.3	About 1.0	5
4	0.3	0.5	1:1	120–300	5.3	About 1.0	5
5	0.3	0.5	1:1	270	5.3 and 8.0	About 1.0	2
6	0.3	0.5	1:1	270	5.3	1.0–15.0	4

2.6. Improved Oil Recovery Effect of Foam Assistant Steam Flooding

2.6.1. Single-Tube Model Experiment

Firstly, the foam-assisted steam flooding efficiency experiment was carried out using a single sand packing model. The experimental flow chart is shown in Figure 1. The experimental steps are as follows: Step (1)–Step (3) are the same as those in Section 2.5. (4) Raise the incubator to the set temperature, keep at a constant temperature for 3 h, and use 3 to 5 times the pore volume of crude oil to displace the water in the core at a rate of 0.5 mL/min so that the oil saturation in the core meets the experimental design requirements. (5) For the steam + PCS flooding experiments, set the incubator at 270 °C for 5 h at a constant temperature, make the back pressure slightly lower than the saturated vapor pressure at this temperature, turn on the pump and inject water (or chemicals) and air into the foam generator according to the designed gas–liquid ratio. Then, open the bypass valve to let the injected mixed fluid pass through the bypass, so that the coil is completely filled with the mixed fluid, and then close the bypass valve. When the inlet pressure of the sand pack is slightly lower than the saturated vapor pressure at the experimental temperature, open the inlet valve and outlet valve of the sand pack, and then start the displacement experiment. (6) The amount of oil and water produced is calculated by solvent extraction distillation method. A total of 4 sets of experiments were carried out: (1) steam flooding; (2) steam + surfactant (CRF); (3) steam + CX-5 + air; (4) steam + PCS + air. During the injection process, the total injection speed was controlled to remain unchanged at 6 mL/min. If the air was injected, the gas–liquid ratio was 1:1, and the back pressure was 4.5 MPa.

2.6.2. Three Parallel Sand Packing Model Displacement Experiment

The heterogeneity of the reservoir is simulated by the parallel sand packing tube model, and the effect of the control and flooding system on the channeling phenomenon during steam flooding and the effect of enhancing oil recovery are studied. The experimental steps are the same as those in Section 2.6.1, only the single sand packing model is replaced with a three-parallel sand packing model. The permeabilities of the three sand packing tubes are set to 1D, 4D, and 10D, respectively, and the model permeability gradient is 10. Four sets of experiments were also carried out: (1) steam flooding; (2) steam + surfactant (CRF); (3) steam + CX-5 + air; (4) steam + PCS + air.

2.6.3. Large-Size Rectangular Sand Filling Mold Displacement Experiment

In order to further study the effect of the PCS in controlling steam overburden, blocking steam channeling channels, and expanding the swept volume of steam flooding, a large-scale sand-filled core insulating air foam steam flooding experiment was carried out. The flow chart of the experimental device and the experimental steps are the same as those in Section 2.6.1.

3. Result and Discussion

3.1. Modified Lignin Properties

3.1.1. FT-IR Spectrum

Figure 2 compares the infrared spectra of the lignin and modified lignin. Figure 2a has a strong hydroxyl absorption peak at $3417\ \mathrm{cm}^{-1}$, which is formed by the normal absorption peaks of hydroxyl at $3650\ \mathrm{cm}^{-1}$ and $3600\ \mathrm{cm}^{-1}$, which are shifted back and strengthened. This is due to the association of the hydroxyl hydrogen bonds on acid or base groups, indicating that lignin contains a large number of hydroxyl groups, and these hydroxyl groups form hydrogen bonds. Peaks at $1329\ \mathrm{cm}^{-1}$, $1219\ \mathrm{cm}^{-1}$, and $1124\ \mathrm{cm}^{-1}$ were mainly produced by syringyl. The peak at $1267\ \mathrm{cm}^{-1}$ belonged to the guaiac group in the lignin structural unit. From the comparison of the peak intensities of the two, it is shown that the content of the guaiac group in lignin is greater than that of the syringyl group. The peaks at $2845\ \mathrm{cm}^{-1}$ and $2938\ \mathrm{cm}^{-1}$ are due to the vibration of the methoxy groups, and the peaks at $3428\ \mathrm{cm}^{-1}$ are masked by the strong and broad absorption due to hydrogen bonding. The other peaks are consistent with the characteristic absorption peaks of chemical functional groups, see Table 2 for details.

Figure 2b still has a strong hydroxyl absorption peak at $3436\ \mathrm{cm}^{-1}$, indicating that there are still many hydroxyl groups that can continue to participate in the reaction after lignin sulfonation. The peak at $2932\ \mathrm{cm}^{-1}$ is due to the methoxy vibration, and at $1061\ \mathrm{cm}^{-1}$ is the coincident peak of the $-SO_3-$ antisymmetric and symmetric bond stretching vibration peaks. By comparing Figure 2a,b, it can be found that a large number of hydroxyl groups in lignin are involved in the oxidation reaction, and sulfonic acid is introduced, which proves the successful synthesis of modified lignin CRF.

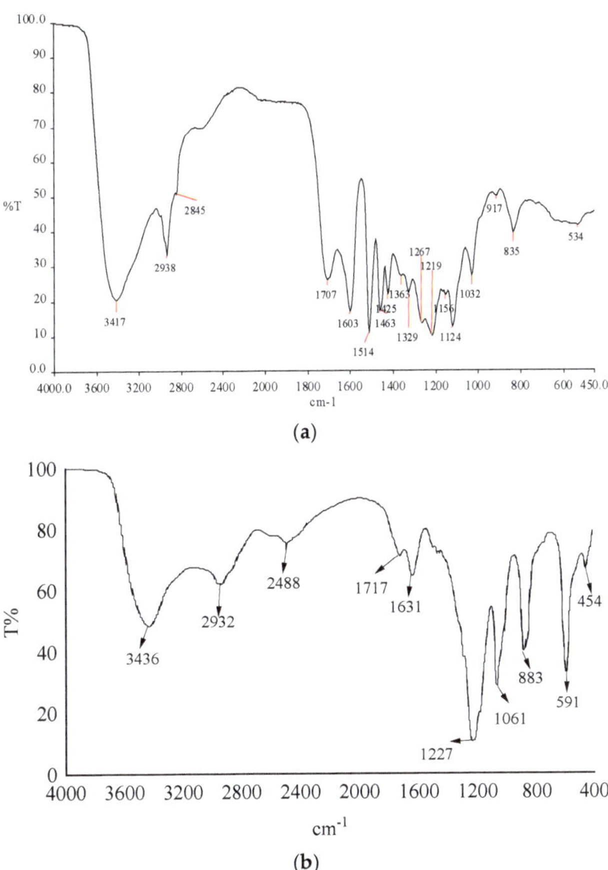

Figure 2. Comparison of infrared spectra of lignin (**a**) and CRF (**b**).

Table 2. The characteristic absorption peak and assignment of infrared spectroscopy.

Wave Number, cm^{-1}	Groups
3417/3436	Stretching vibration characteristic absorption peaks of hydroxyl O-H
2938, 2845	The characteristic absorption peak of methyl stretching vibration
1707	Characteristic absorption peaks of non-conjugated carbonyl groups
1603	Aromatic ring skeleton vibration absorption peak
1514	Aromatic ring C-H bending vibration
1400~1460	Characteristic absorption peaks of stretching vibration of C-H of methylene
1329	Vibrational absorption peaks of lilac-type benzene ring skeleton
1267	Guaiac wood ring (C-O)
1219	C-O stretching vibration of the syringyl ring
1124	Syringyl group
1061	-SO$_3$- antisymmetric and symmetric bond stretching vibration peaks coincide
1032	Secondary alcohol or ether (C-H bending vibration)
835	Out-of-plane C-H vibrations of aromatic rings at the 2, 5, and 6 positions (guaiacol type)

3.1.2. UV Spectral Comparison

Figure 3 compares the UV spectra of lignin and modified lignin. the lignin and CRF have absorption peaks around 212.50 nm and 202 nm, respectively, which are the absorption bands of conjugated olefinic bonds. A broad shoulder peak appears between 250 nm and

300 nm, indicating that there is a conjugated system in the molecule, which is the absorption band of the aromatic rings. The weak absorption peak around 278 nm indicates that the two have more symmetrical syringylbenzene ring structures. The UV spectrum shows that the CRF is still dominated by the aromatic structure of lignin, but sulfonate is introduced into the side chain through sulfonation. The benzene ring structure provides CRF with a strong temperature resistance.

Figure 3. Comparison of UV spectra of lignin (**a**) and CRF (**b**).

3.1.3. Elemental Analysis

From the elemental analysis in Table 3, it can be concluded that the molecular formula of CRF can be expressed as $C_{95}H_{128}O_{41}S_4$. The high C/H atomic ratio indicates that the lignin has a high content of benzene rings and a large proportion of oxygen, indicating that the lignin contains more hydroxyl groups, methoxy groups, and ether bonds.

Table 3. The element analysis results of lignin.

Elements	C	H	O	S	C/H
Contents, %	55.55	6.28	31.93	6.24	8.81

It can be seen from the ultraviolet spectrum, infrared spectrum, and elements analysis that the modified lignin molecules mainly include guaiac groups and syringyl groups, of which syringyl groups are more than lignin groups. And the sulfonic acid group was introduced through oxidative modification, so the molecular chemical formula of CRF can be simulated, as shown in Figure 4.

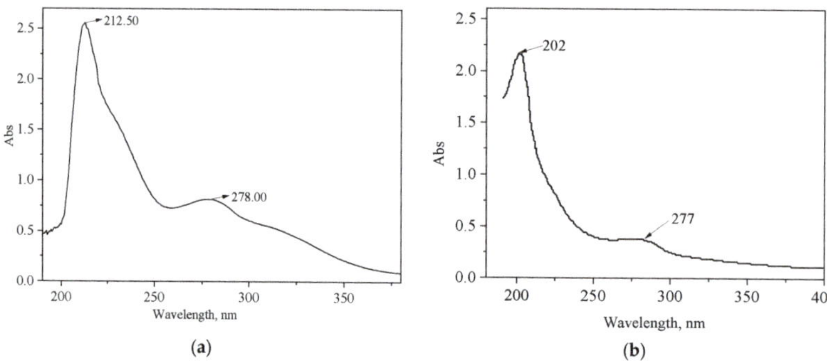

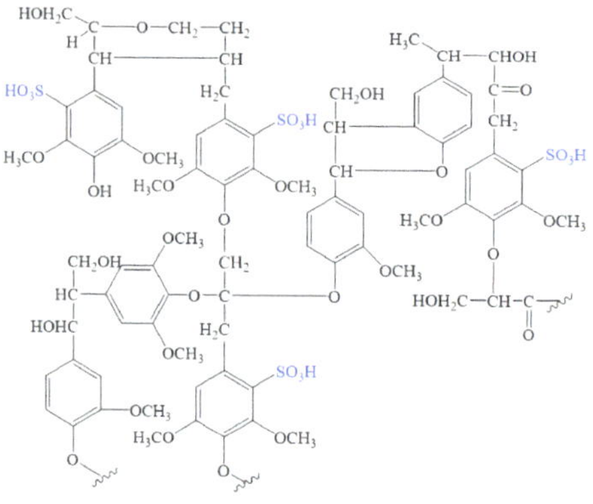

Figure 4. The simulated molecular formula of CRF.

3.1.4. Thermogravimetric Analysis

Figure 5 shows the thermogravimetric (TG) and differential thermogravimetric (DTG) curves of CRF at a heating rate of 15 °C/min. The obvious mass loss of CRF at 75~160 °C was caused by the desorption of free water, and the mass loss was 3.8%. The maximum mass loss rate was 0.75%/min, and the corresponding peak temperature was 140 °C. The main pyrolysis range of CRF is 235~415 °C, the mass loss is 58.58%, the maximum mass loss rate is 5.5%/min, and the corresponding peak temperature is 376.3 °C. The final yield of solid residue after pyrolysis was 39.8%, indicating that CRF has good temperature resistance in this temperature range.

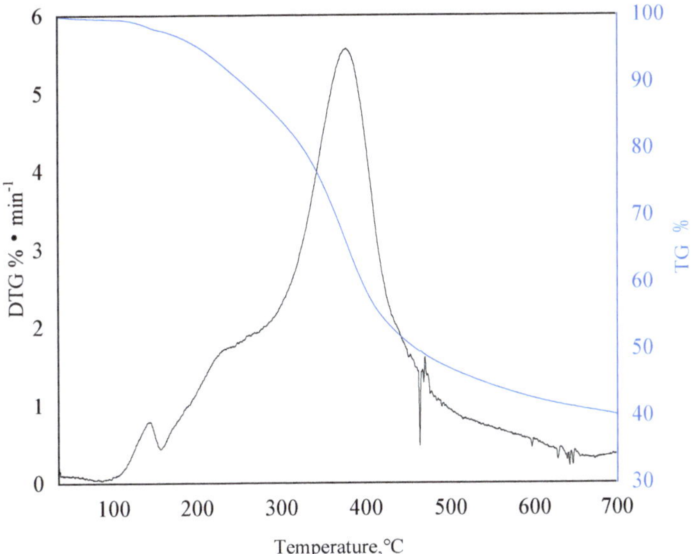

Figure 5. The TG curve and DTG curve of lignin (15 °C/min).

3.2. Foaming Performance of CX-5

Figure 6 shows the foaming volume and half-life curves of CX-5 at the three temperatures 20 °C, 200 °C, and 300 °C. Figure 6a shows that at 20 °C, the foaming volume increases first and then tends to be stable with the increase in concentration. After the temperature was raised to 200 and 300 °C, the foaming volume of CX-5 slightly decreased, and the foaming volume with different CX-5 concentrations was similar and higher than 180 mL, showing good foaming performance. Figure 6b shows that the half-life of the foam decreases slightly after the temperature increases from 20 °C to 200 °C, but the half-life remains basically unchanged when the temperature continues to increase to 300 °C, showing good temperature resistance. When the concentration of CX-5 is 0.1 wt% at the three temperatures, the foam stability is the strongest; if the concentration of CX-5 continues to increase, the stability of the foam becomes worse. This is mainly because, when the concentration is higher than the critical micelle concentration, the enrichment of surfactant molecules on the surface of the solution reduces the liquid content and the stability of the foam. The half-life of CX-5 is more than 450 s as a whole, and it has good foam stability. Taking into account the foaming volume and half-life of the foam, the concentration range of CX-5 should be between 0.1 wt% and 0.3 wt%.

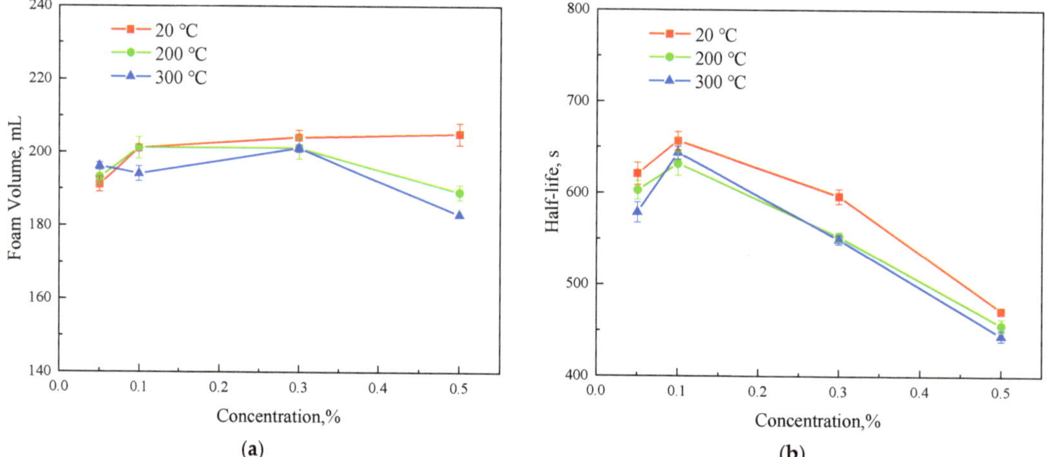

Figure 6. The influence of temperature on the foaming performance of CX-5. (**a**) Foam volume, (**b**) foam stability.

3.3. Surface Tension of CX-5

The foam system has a large specific surface area. The lower the surface tension, the more conducive to the stability of the foam, the less energy input from the outside is required, and the more conducive to the generation of foam. Therefore, surface tension is an important property of the foam system. Figure 7 shows the effect of temperature and concentration factors on surface tension.

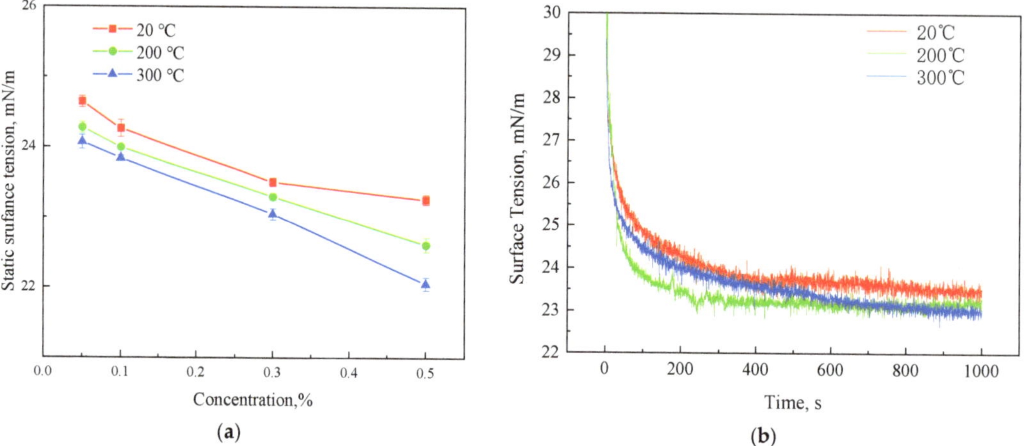

Figure 7. The influence of temperature on the surface tension of CX-5. (**a**) Static surface tension, (**b**) dynamic surface tension.

Figure 7a shows that when the temperature is 20 °C, the surface tension of the solution first decreases and then tends to be stable with the increase in the concentration of CX-5; when the temperature increases to 200 °C and 300 °C, the surface tension of the solution continued to decrease with the increase in CX-5 concentration. At the same concentration, the surface tension of the solution decreases gradually with the increase in temperature. This is mainly because, as the temperature increases, the molecular thermal motion intensifies, which reduces the surface tension. Meanwhile, because CX-5 has good temperature resistance, it can still effectively reduce the interfacial tension at high temperatures. Figure 7b shows the change in the dynamic surface tension of the CX-5 solution. It can be found that the surface tension decreases rapidly during 0–200 s with the injection of gas, decreases slowly during 200–600 s, and stabilizes after 600 s. Figure 7 shows that the CX-5 solution has good surface activity, which can rapidly reduce the surface tension and maintain stability under high-temperature conditions, which is the basis for its good foaming performance.

3.4. Resistance Factor of the Foam

3.4.1. Formulation Optimization of PCS

Like polymer flooding, enhanced flow resistance is the basic EOR mechanism [42]. The two main agents of the CRF solution are the oil displacement agent CRF and foaming agent CX-5, and the recommended concentration of both is 0.2~0.5%, according to the static performance evaluation. In order to further clarify the optimal concentration of CRF and CX-5 in the PCS solution, resistance factor tests were carried out according to the experimental scheme in Table 1 (#1 and #2). The resistance factors of the different concentrations of CRF and CX-5 are shown in Figure 8.

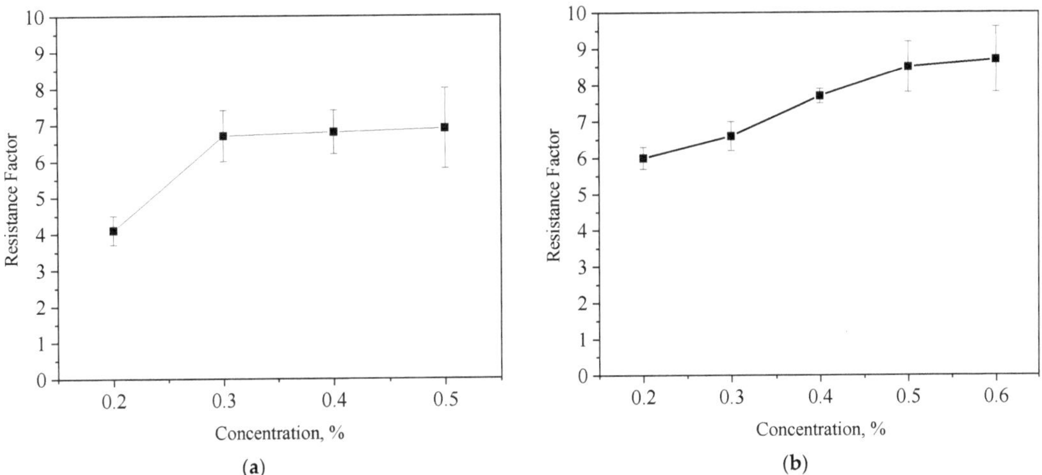

Figure 8. Resistance factor curves. (**a**) Different concentration of CRF (CX-5 is 0.3 wt%); (**b**) different concentration of CX-5 (CRF is 0.3 wt%).

It can be seen from Figure 8a that, when the concentration of CX-5 is constant (0.3 wt%), with the increase in CRF concentration, the resistance factor gradually increases, but when the concentration reaches 0.3 wt%, the increase decreases significantly. So, we set the concentration of CRF to 0.3 wt%. Figure 8b shows that when the CRF concentration is constant (0.3%), the resistance factor gradually increases with the increase in the CX-5 concentration, but, when the concentration reaches 0.5%, the increase decreases significantly. So, the concentration of CX-5 was set at 0.5 wt%. Finally, the formulation of the PCS solution for steam flooding in heavy oil reservoirs determined in this paper is 0.3 wt% CRF + 0.5 wt% CX-5.

3.4.2. Influence of Gas–Liquid Ratio on Resistance Factor of PCS

Table 4 shows the calculation results of the resistance factor of the PCS solution when the gas–liquid ratio is 0:1, 1:2, 1:1, 3:2, and 2:1. When the gas-to-liquid ratio is too high or too low, it is unfavorable for the foam system to play a mobility control role in porous media. The gas–liquid ratio has a high resistance factor in the range of 1:2~2:1. Therefore, the gas–liquid ratio should be controlled as much as possible between 1:2~2:1 in the field construction.

Table 4. The influence of different gas–liquid ratios on the resistance coefficient.

Gas–Liquid Ratio	Permeability, D	Basic Pressure, Mpa	Work Pressure, MPa	Resistance Factor
0:1	1.05	0.0104	0.0718	6.9
1:2	1.07	0.0107	0.0813	7.0
1:1	1.07	0.0106	0.0901	8.5
2:1	1.09	0.0112	0.0833	7.5
3:2	1.06	0.0105	0.0704	6.7

3.4.3. Influence of Temperature on Resistance Factor of PCS

Table 5 shows the resistance factors of PCS solution under the conditions of 120 °C, 150 °C, 200 °C, 270 °C, and 300 °C. As the temperature increases, the resistance factor decreases. When the temperature is 270 °C, the resistance factor is 8.4, and when the experimental temperature reaches 300 °C, the resistance factor is 5.8. In the process of steam injection, when the resistance factor is greater than 4, the injected flooding agent plays the role of diverting the steam. Therefore, when the temperature exceeds 270 °C, the injected PCS solution still has the effect of improving the displacement profile.

Table 5. The influence of temperature on the resistance factor.

Temperature, °C	Permeability, D	Basic Pressure, MPa	Work Pressure, MPa	Resistance Factor
120	1.11	0.0099	0.4990	50.4
150	1.13	0.0103	0.2029	19.7
200	1.08	0.0096	0.1123	11.7
270	1.07	0.0095	0.0796	8.4
300	1.09	0.0097	0.0563	5.8

3.4.4. Influence of Pressure on Resistance Factor of PCS

In order to evaluate the effect of pressure on the foam plugging ability, foam plugging experiments were carried out under the back pressure of 5.3 MPa and 8 MPa, respectively. The calculation results of the resistance factor are shown in Table 6. The increase in pressure can increase the stability of the foam, and, correspondingly, increase the plugging ability of the foam. As the pressure increases, the diameter of the foam becomes smaller, which is beneficial to the stability of the foam. After the pressure was increased from 5.3 MPa to 8 MPa, the resistance factor increased from 8.4 to 10.5.

Table 6. The influence of return pressure on the resistance factor.

Back Pressure, MPa	Permeability, D	Basic Pressure, Mpa	Work Pressure, MPa	Resistance Factor
5.3	1.07	0.0095	0.0796	8.4
8	1.09	0.0103	0.1079	10.5

3.4.5. Influence of Permeability on Resistance Factor of PCS

The sand packs with the permeability of 1D, 4D, 10D, and 15D were used to evaluate the foam resistance factor, and the experimental results are shown in Table 7. The results show that with the increase in permeability, the resistance factor increases correspondingly, which shows that the foaming ability of air foam in the high-permeability region with large pores is higher than that in the low-permeability region with small pores. This is one of the important bases for using air foam to plug a high-permeability layer and improve the sweep coefficient.

Table 7. The influence of permeability on the resistance factor.

Permeability, D	Basic Pressure, Mpa	Work Pressure, MPa	Resistance Factor
1.07	0.0105	0.0796	8.4
4.38	0.0082	0.0812	9.9
9.80	0.0063	0.0706	11.2
15.2	0.0042	0.0487	11.6

3.5. Enhance Oil Recovery Effect of PCS Assistant Steam Flooding

3.5.1. Single Sand Pack Flooding Experiments

Figures 9 and 10 are the oil displacement characteristic curves of the four displacement modes at 200 °C and 270 °C, respectively. The steam flooding efficiency can be increased from 78% to 81.2% after the temperature is increased from 200 °C to 270 °C. Under the two temperature conditions, both CRF-assisted steam flooding and CX-5-assisted steam flooding can improve the oil displacement efficiency on the basis of steam flooding. Compared with thermal steam flooding, the oil displacement efficiency is increased by 3.5%, 1.7%, and 7.9%, respectively at 270 °C. When used alone, the oil displacement efficiency improvement of CRF is higher than that of CX-5. The combined use of the two can further improve the recovery factor by more than 6%, and the effect of PCS is good.

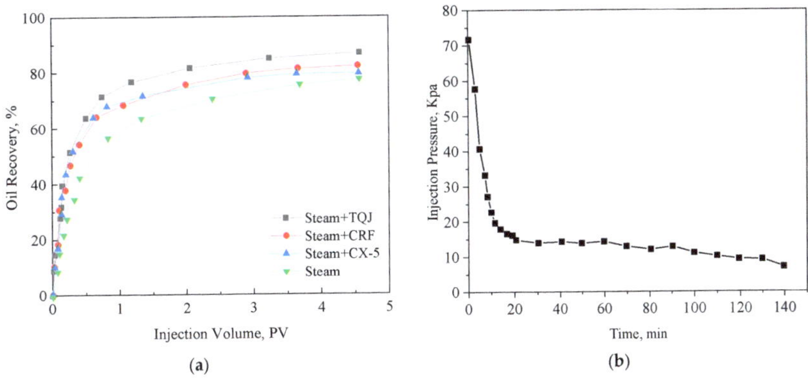

Figure 9. Oil displacement curves at 200 °C. (**a**) Oil displacement efficiency, and (**b**) the pressure differential curve of steam flooding.

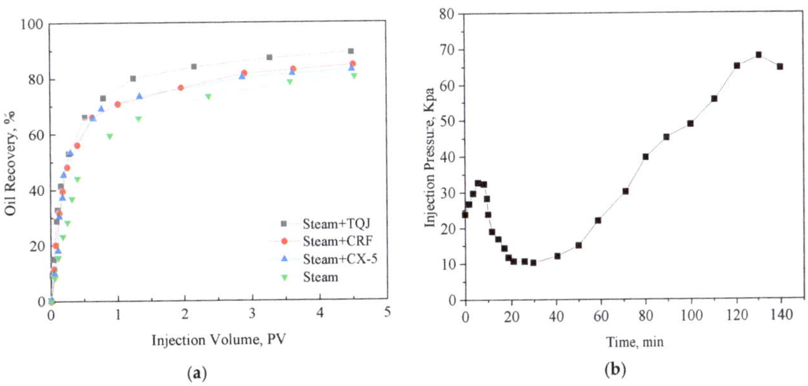

Figure 10. Oil displacement curves at 270 °C. (**a**) Oil displacement efficiency, (**b**) the pressure differential curve of TQJ assistant steam flooding.

When the PCS solution is added to the hot water zone and the steam zone for displacement, the oil saturation in the sand pack gradually decreases as the displacement progresses, and the foam gradually plays a plugging role. From the pressure difference curve, it can be seen that the pressure difference between the two ends of the sand pack gradually increases in the middle and late stages of displacement (Figures 9b and 10b), which is equivalent to a decrease in the steam mobility by up to eight times.

3.5.2. Three Parallel Sand Packs Flooding Experiments

Figure 11 is the characteristic curve of the three parallel flooding experiments. During steam flooding, due to the serious formation heterogeneity, the sweep efficiency of the lower part of the oil layer is poor. When the flooding experiment is completed (the water cut of the produced liquid is 98%), the remaining oil saturation in the three sand packs of the low-, medium-, and high-permeability layers are 48.8%, 24.9%, and 12.0%, respectively, and the total oil recovery is 54.5%. The CRF assistant steam flooding has a certain profile control ability, and the swept condition of the low-permeability layer is improved. The residual oil saturation decreased by 10.7%. The oil saturation of the medium-permeability layer has little change, and the oil recovery is also improved compared with steam flooding, which is increased by 7.9%. Although the effect of steam flooding with CX-5 is significantly better than that of steam flooding with CRF, the residual oil in the low-permeability zone is still as high as 22.7% at the end of the test. The ultimate recovery factor is further increased by 12%. Finally, the sweeping condition of steam flooding with PCS has been greatly improved. It not only improves the displacement effect of the low-permeability layer by 52.3% but also improves the displacement effect of the medium- and high-permeability layers, and the total oil recovery is 28.7% higher than that of steam flooding.

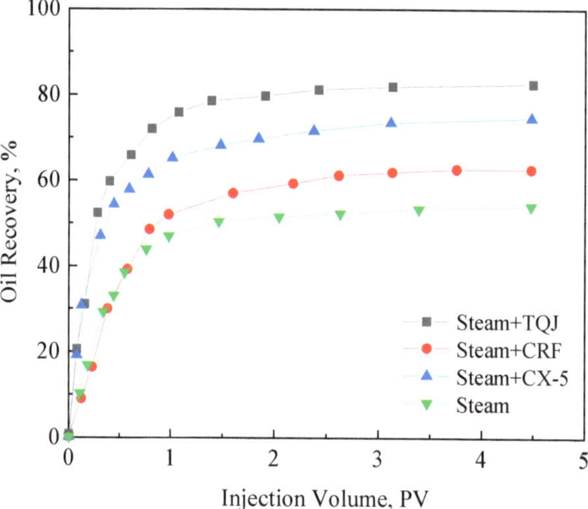

Figure 11. The curve of oil displacement efficiency and amount of injection PV in three tubes displacement experiments.

The differential pressure curves of steam flooding with and without PCS are shown in Figure 12. Consistent with the experimental results of the single tube flooding experiments, the injection–production pressure difference increases significantly after the injection of PCS, which has a strong effect on the expansion of the swept volume of the heterogeneous oil reservoir.

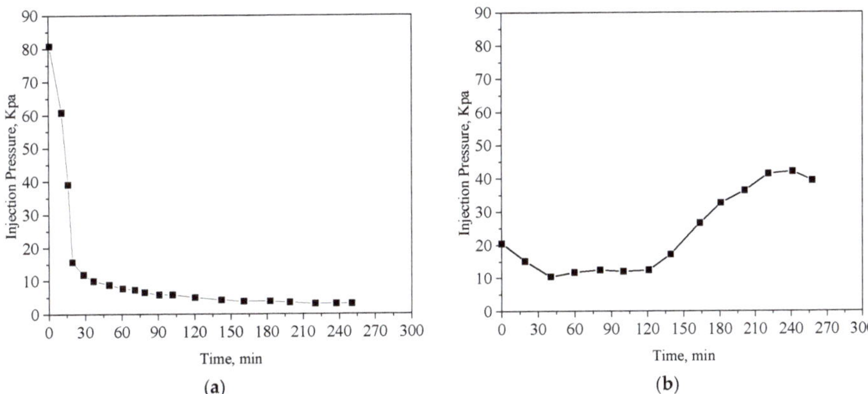

Figure 12. The pressure differential curve of sand-filled pipe core. (**a**) The steam flooding process, and (**b**) the PCS-assisted steam flooding process.

3.5.3. Large-Size Rectangular Sand Filling Model Flooding Experiments

During the experiment, steam flooding was carried out first. When the water cut of the produced fluid reached 98%, a 0.5 PV PCS solution was injected, and then continued steam flooding. When the water cut reached 98%, the experiment was terminated. Figure 13 shows the oil displacement efficiency curves of the homogeneous cores and heterogeneous cores. Figure 13 shows that for a large-sized homogeneous core, when the water cut reaches 98.0%, the oil displacement efficiency of conventional steam flooding is 57.53%. And the oil displacement efficiency is increased by 18.4% after the PCS flooding and second steam flooding. For the large-sized heterogeneous cores, when the water cut reaches 98.0%, the oil displacement efficiency of conventional steam flooding is 44.83%. After PCS flooding and the second steam flooding, the oil displacement efficiency increased by 27.86%. Comparing the results of the homogeneous and heterogeneous model flooding experiments, it can be found that the oil displacement efficiency is significantly improved when the water cut reaches 98.0% after steam flooding, indicating that the sweep efficiency of the homogeneous and heterogeneous cores has been improved. But PCS has a more obvious effect on the heterogeneous core, and its oil displacement efficiency increment is 9.46% higher than that of the homogeneous core.

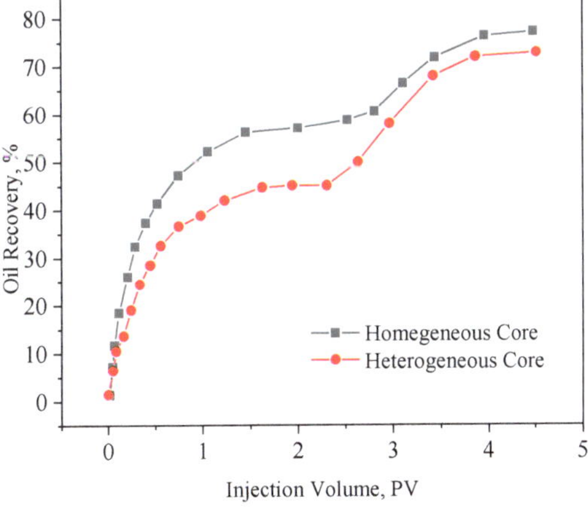

Figure 13. The curve of oil displacement efficiency and the number of injected PV.

4. Conclusions

In this paper, a high-temperature-resistant modified lignin, CRF, was developed and compounded with a foaming agent, CX-5, as a steam flooding profile control system (PCS). This paper systematically evaluates the foaming performance, resistance factor, and influencing factors of CX-5 and finally evaluates the enhanced oil recovery effect of the PCS-assisted steam flooding. The specific conclusions are as follows:

(1) CRF can be synthesized by sulfonation and the introduction of nonylphenol into the side chain of lignin molecules after oxidative chain scission. The FT-IR, UV-visible, and elemental analysis showed that CRF was successfully synthesized. CRF contains mostly syringyl groups, followed by guaiac groups, explaining its good temperature resistance. Low molecular weight, homogenization, and high activity are the directions for the development of lignin modification.

(2) CX-5 is a commercially modified anionic surfactant with a significantly increased molecular weight, giving it high-temperature resistance and interface activity. Its optimal concentration range is 0.3 wt%~0.5 wt%, the foaming volume currently is greater than 180 mL, and the foam half-life is greater than 400 s.

(3) The optimal formulation of PCS was determined by the resistance factors: 0.3 wt% CRF as an oil displacement agent + 0.5 wt% CX-5 as a foaming agent. The gas–liquid ratio is 1:2 to 2:1, and the formula system can withstand temperatures above 270 °C.

(4) The extent of reducing the residual oil saturation by PCS decreases with the increase in temperature. When the temperature is 200 °C, the oil displacement efficiencies of steam flooding with CRF, CX-5, and PCS are 5%, 2.1%, and 9% higher than that of steam flooding, respectively; and at 270 °C, the oil displacement efficiencies are 3.5%, 1.7%, and 7.9% higher than that of steam flooding, respectively. PCS can increase the injection–production pressure difference to more than eight times that of steam flooding, which can effectively expand the swept volume of steam.

(5) The three parallel tubes flooding experiment shows that the total oil recovery of PCS-assisted steam flooding increased by 28.7% based on steam flooding reaching 83.2%. PCS-assisted steam flooding can effectively improve the sweep efficiency of heterogeneous cores, which is the basic mechanism of EOR, resulting in a more than 9.46% EOR than the homogeneous cores. Our next research target is to release the microscopic pore throat flow rules and EOR mechanism of PCS-assisted steam flooding.

Author Contributions: Conceptualization, L.D.; Methodology, Q.H. and F.M.; Formal analysis, F.Z. and H.Z.; Data curation, Y.L.; Writing—original draft, D.L.; Writing—review & editing, S.G.; Funding acquisition, L.D. All authors have read and agreed to the published version of the manuscript.

Funding: This research was funded by [the Natural Science Foundation of Heilongjiang Province] grant number [LH2023B014].

Institutional Review Board Statement: Not applicable.

Data Availability Statement: Data are contained within the article.

Acknowledgments: The research is supported by the Natural Science Foundation of Heilongjiang Province (LH2023B014). The authors also thank the anonymous reviewers for their valuable comments.

Conflicts of Interest: The authors declare no conflict of interest.

References

1. Lyu, X.; Liu, H.; Pang, Z.; Sun, Z. Visualized study of thermochemistry assisted steam flooding to improve oil recovery in heavy oil reservoir with glass micromodels. *Fuel* **2018**, *218*, 118–126. [CrossRef]
2. Zhao, D.W.; Wang, J.; Gates, I.D. Optimized solvent-aided steam-flooding strategy for recovery of thin heavy oil reservoirs. *Fuel* **2013**, *112*, 50–59. [CrossRef]
3. Pang, Z.; Wang, L.; Yin, F.; Lyu, X. Steam chamber expanding processes and bottom water invading characteristics during steam flooding in heavy oil reservoirs. *Energy* **2021**, *234*, 121214. [CrossRef]
4. Palatz. *Thermal Oil Recovery*; Petroleum Industry Press: Beijing, China, 1989; pp. 13–16.
5. Chen, Y. *Steam Injection Thermal Oil Recovery*; Petroleum University Press: Dongying, China, 1996.

6. Pang, Z.; Lyu, X.; Zhang, F.; Wu, T.; Gao, Z.; Geng, Z.; Luo, C. The macroscopic and microscopic analysis on the performance of steam foams during thermal recovery in heavy oil reservoirs. *Fuel* **2018**, *233*, 166–176. [CrossRef]
7. Shen, P. *Polymer Flooding Enhanced Oil Recovery Technology*; Petroleum Industry Press: Beijing, China, 2006; pp. 1–2.
8. Zhao, F. Special Profile Control Agent for Thermal Recovery of Heavy Oil Wells. CN101423753A, 23 September 2008.
9. Wright, T.R. Initial downhole steam generator tests completed. *World Oil* **1980**, *19*, 15.
10. Wright, T.R. Comparative analysis of steam delivery cost for surface and downhole steam drive technologies. *World Oil* **1981**, *16*, 11–12.
11. Gao, S.; Li, H.; Yan, Z.; Pitts, M.J.; Surkalo, H.; Wyatt, K. The alkaline/surfactant/polymer pilot performance of search, west central, Daqing oil field. *SPE Reserv. Eng.* **1996**, *11*, 181–188.
12. Hu, C. *Heavy Oil Recovery Technology*; Petroleum Industry Press: Beijing, China, 1999; pp. 23–24.
13. Li, Y.; Chen, X.; Liu, Z.; Liu, R.; Liu, W.; Zhang, H. Effects of molecular structure of polymeric surfactant on its physicochemical properties, percolation and enhanced oil recovery. *J. Ind. Eng. Chem.* **2021**, *101*, 165–177. [CrossRef]
14. Chen, X.; Li, Y.; Liu, Z.; Zhang, J.; Chen, C.; Ma, M. Investigation on matching relationship and plugging mechanism of self-adaptive micro-gel (SMG) as a profile control and oil displacement agent. *Powder Technol.* **2020**, *364*, 774–784. [CrossRef]
15. Chen, X.; Li, Y.; Liu, Z.; Li, X.; Zhang, J.; Zhang, H. Core-and pore-scale investigation on the migration and plugging of polymer microspheres in a heterogeneous porous media. *J. Pet. Sci. Eng.* **2020**, *195*, 107636. [CrossRef]
16. Chen, X.; Li, Y.; Liu, Z.; Zhang, J.; Trivedi, J.; Li, X. Experimental and theoretical investigation of the migration and plugging of the particle in porous media based on elastic properties. *Fuel* **2023**, *332*, 126224. [CrossRef]
17. Chen, X.; Li, Y.; Liu, Z.; Trivedi, J.; Tang, Y.; Sui, M. Visualized investigation of the immiscible displacement: Influencing factors, improved method, and EOR effect. *Fuel* **2023**, *331*, 125841. [CrossRef]
18. Kong, D.; Gao, J.; Lian, P.; Zheng, R.; Zhu, W.; Xu, Y. Characteristics of gas-oil contact and mobilization limit during gas-assisted gravity drainage process. *Adv. Geo-Energy Res.* **2022**, *6*, 169–176. [CrossRef]
19. Kong, D.; Gao, Y.; Sarma, H.; Li, Y.; Guo, H.; Zhu, W. Experimental investigation of immiscible water-alternating-gas injection in ultra-high water-cut stage reservoir. *Adv. Geo-Energy Res.* **2021**, *5*, 139–152. [CrossRef]
20. Liu, P.; Zhang, X.; Wu, Y.; Li, X. Enhanced oil recovery by air-foam flooding system in tight oil reservoirs: Study on the profile-controlling mechanisms. *J. Pet. Sci. Eng.* **2017**, *150*, 208–216. [CrossRef]
21. Lang, L.; Li, H.; Wang, X.; Liu, N. Experimental study and field demonstration of air-foam flooding for heavy oil EOR. *J. Pet. Sci. Eng.* **2020**, *185*, 106659. [CrossRef]
22. Chen, X.; Li, Y.; Sun, X.; Liu, Z.; Liu, J.; Liu, S. Investigation of Polymer-Assisted CO_2 Flooding to Enhance Oil Recovery in Low-Permeability Reservoirs. *Polymers* **2023**, *15*, 3886. [CrossRef]
23. Needham, R.B. Plugging High Permeability Earth Strata. U.S. Patent No 3412793, 26 November 1968.
24. Paul, G.W.; Lake, L.W.; Pope, G.A.; Young, G.B. A simplified predictive model for micellar-polymer flooding. In Proceedings of the SPE California Regional Meeting, San Francisco, CA, USA, 24 March 1982.
25. Scheren, M. Parameter influencing porosity in sandstone a model for sandstone porosity prediction. *AAPG Bull.* **1987**, *71*, 485–491.
26. Campbell, T.C. The role of alkaline chemicals in the recovery of low-gravity crude oils. *J. Pet. Technol.* **1982**, *34*, 2510–2516. [CrossRef]
27. Duerksen, J.H.; Hsueh, L. Steam distillation of crude oil. *SPE J.* **1983**, *23*, 265–271. [CrossRef]
28. Gorgarty, W.B. Mobility Control with polymer Solutions. *SPE J.* **1967**, *6*, 161–173.
29. Smith, J.E. Quantitative Evaluation of Polyacrylamide Crosslinked Gels for Enhanced Oil Recovery. In Proceedings of the International ACS Symposium, Anaheim, CA, USA, 18 June 1986; pp. 9–12.
30. Clampitt, R.L. Selective Plugging of Formations with Foam. U.S. Patent No 3993133, 23 November 1976.
31. Dilgren, R.E. Stram-Channel-Expand Steam Foam Drive. U.S. Patent No 4086964, 2 May 1978.
32. Dilgren, R.E.; Owens, K.B. Hot Water Foam Oil Production Process. U.S. Patent No 4161217, 17 July 1979.
33. Borchard, J.K.; Bright, D.B.; Dickson, M.K.; Wellington, S.L. Surfactants for CO_2 Foam Flooding. In Proceedings of the SPE14394, Las Vegas, NV, USA, 22–26 September 1985.
34. Ettinger, R.A.; Radke, C.J. Influence of texture oil steady foam flow in Berea sandstone. *SPE Reserv. Eng.* **1992**, *7*, 169–268. [CrossRef]
35. Schramm, L.L.; Turta, A.T.; Novosad, J.J. Microvisual and core flood studies of foam Interactions with a light crude oil. *SPE Reserv. Eng.* **1993**, *8*, 245–257. [CrossRef]
36. Majeed, T.; Kamal, M.S.; Zhou, X.; Solling, T. A review on foam stabilizers for enhanced oil recovery. *Energy Fuels* **2021**, *35*, 5594–5612. [CrossRef]
37. Zhang, Y.; Liu, Q.; Ye, H.; Yang, L.; Luo, D.; Peng, B. Nanoparticles as foam stabilizer: Mechanism, control parameters and application in foam flooding for enhanced oil recovery. *J. Pet. Sci. Eng.* **2021**, *202*, 108561. [CrossRef]
38. Bhatt, S.; Saraf, S.; Bera, A. Perspectives of Foam Generation Techniques and Future Directions of Nanoparticle-Stabilized CO_2 Foam for Enhanced Oil Recovery. *Energy Fuels* **2023**, *37*, 1472–1494. [CrossRef]
39. Abdelaal, A.; Gajbhiye, R.; Al-Shehri, D. Mixed CO_2/N_2 foam for EOR as a novel solution for supercritical CO_2 foam challenges in sandstone reservoirs. *ACS Omega* **2020**, *5*, 33140–33150. [CrossRef]
40. Richardson, E.A.; Scheuerman, R.F. Plugging Solution Precipitation Time Control. U.S. 3730272, 1 May 1973.

41. Wang, M. Thermal oil recovery and enhanced oil recovery. *Oil Gas Recovery Technol.* **1994**, *1*, 6–11.
42. Dong, L.; Li, Y.; Wen, J.; Gao, W.; Tian, Y.; Deng, Q.; Liu, Z. Functional characteristics and dominant enhanced oil recovery mechanism of polymeric surfactant. *J. Mol. Liq.* **2022**, *354*, 118921. [CrossRef]

Communication

Synthesis and Performance Testing of Maleic Anhydride–Ene Monomers Multicomponent Co-Polymers as Pour Point Depressant for Crude Oil

Dong Yuan [1,*], Qingfeng Liu [2], Wenhui Zhang [3], Ran Liu [1,3], Chenxi Jiang [3], Hengyu Chen [3], Jingen Yan [4], Yongtao Gu [4] and Bingchuan Yang [1,3,*]

1   College of Chemistry and Chemical Engineering, Qilu Normal University, Jinan 250013, China; liuranbjt@163.com
2   Shengli Oil Field Zhongsheng Petroleum Development Company, Dongying 257237, China; wolfmanliu007@163.com
3   School of Chemistry and Chemical Engineering, Liaocheng University, Liaocheng 252000, China; zwh2963105101@163.com (W.Z.); jcx20010331@163.com (C.J.); qicechenyu@163.com (H.C.)
4   Gudong Oil Production Plant, Shengli Oilfield Company, SINOPEC, Dongying 257237, China; yanjingen@163.com (J.Y.); guyongtao@163.com (Y.G.)
*   Correspondence: yuandong@mail.sdu.edu.cn (D.Y.); yangbingchuan@lcu.edu.cn (B.Y.)

Abstract: To address the issue of pipeline blockage caused by the formation of waxy deposits inside pipelines, hindering the flow of petroleum in the Shengli oilfield, eight new-style polyacrylic acid pour point depressants (PPD) for Shengli crude oil were prepared by maleic anhydride and ene monomers with different polar and aromatic pendant chains. The synthesized Pour Point Depressants were characterized by Fourier transform infrared spectroscopy (FTIR), nuclear magnetic resonance (NMR), gel permeation chromatography (GPC), and polarizing optical microscopy (POM). The results were promising and demonstrated that any type of pour point depressant exhibited excellent performance on high-pour-point crude oil. The reduction in pour-point after additive addition was largely dependent on the polymer structure. Notably, polymers containing long alkyl side chains and aromatic units displayed the most impressive performance, capable of depressing the pour point by 12 °C.

Keywords: crude oil; pour point depressant; maleic anhydride; ene monomers; pour point depressant mechanism

Citation: Yuan, D.; Liu, Q.; Zhang, W.; Liu, R.; Jiang, C.; Chen, H.; Yan, J.; Gu, Y.; Yang, B. Synthesis and Performance Testing of Maleic Anhydride–Ene Monomers Multicomponent Co-Polymers as Pour Point Depressant for Crude Oil. *Polymers* **2023**, *15*, 3898. https://doi.org/10.3390/polym15193898

Academic Editors: Japan Trivedi, Zheyu Liu, Xin Chen, Jianbin Liu and Zhe Li

Received: 19 August 2023
Revised: 18 September 2023
Accepted: 22 September 2023
Published: 27 September 2023

1. Introduction

The Shengli oilfield is located in the northeast region of China, which is a typical cold climate zone with the lowest temperature dropping to −10 °C, and the average winter temperature ranging between 2 and −3 °C [1–3]. Particularly during winter, the extremely low temperatures facilitate the crystallization of wax present in crude oil, leading to the formation of waxy substances that precipitate within the pipeline system and consequently elevate the viscosity of the crude oil [4–7]. According to surveys [8,9], the wax content in Shengli crude oil is relatively high, generally around 10%, and in some areas, it can even reach more than 25%. This makes it extremely difficult to transport crude oil at low temperatures. China is one of the world's largest oil-consuming countries [10], and so, it is urgent to solve the problem of low-temperature transportation of crude oil.

The pour point is the temperature at which crude oil becomes viscous and no longer flows under certain conditions [11]. Above the pour point, crude oil can still flow, while below the pour point, crude oil becomes viscous and difficult to flow [12]. In a low-temperature environment, the pour point of crude oil increases, causing it to solidify and crystallize, making it difficult to flow [13]. Therefore, the pour point is an important indicator of crude oil that needs to be tested and controlled to address the challenges associated with the low-temperature transportation of crude oil [14].

The fundamental way to solve this problem is to add pour point depressants to reduce the pour point of crude oil [15–18]. Crude oil pour point depressants are usually high molecular weight polymers containing a large number of polar functional groups such as hydroxyl and carboxyl groups [19,20]. By means of van der Waals forces, particularly hydrophobic interactions, the polymer interacts with the hydrocarbons in crude oil, forming distinct associations, in which the hydrophilic moieties are oriented on the outside, preventing the agglutination and subsequent precipitation of the waxy molecules [21,22]. There are three mature viewpoints on the principle of crude oil pour point depressants: crystal nucleation theory, adsorption theory, and eutectic theory [23–26]. These theories effectively explain the role of crude oil pour point depressants in improving the flowability of crude oil, laying a foundation for the subsequent preparation and development of pour point depressants [27–29].

Recently, researchers have been gradually exploring the connection between the structure and mechanism of pour point depressants, leading to the development of a wider range of pour point depressants [30–32]. These include (poly)methacrylate-based pour point depressants with improved shear resistance and a variety of maleic anhydride-based pour point depressants with diverse functionalities. Wu et al. synthesized four PPDs with maleic anhydride and its derivatives of different polar and aromatic pendant chains, which are binary copolymers [33]. Fang et al. prepared the ternary copolymers by mixing the aminated copolymer and the composite commercial ethylene-vinyl acetate copolymers (EVA) [34]. Based on previous work, we synthesized eight new-style pour point depressants by polymerizing maleic anhydride and various ene monomers, including acrylic acid and acrylic ester. We tested the effectiveness of these pour point depressants using Shengli crude oil with a pour point of 32 °C and the results demonstrated that any type of pour point depressant exhibited excellent performance on high-pour-point crude oil. Additionally, we explored the mechanism of action of the PPDs and discussed the impact of different functional groups and concentrations on their performance.

2. Materials and Methods

2.1. Materials

Acrylic acid, methyl acrylate, propyl acrylate, 2-butene-1-methyl acrylate, maleic anhydride, methyl methacrylate, vinyl acetate, octadecyl acrylate, nonadecyl acrylate, benzyl alcohol, anhydrous ethanol, benzyl alcohol, diethyl maleate, dibenzyl maleate, sulfuric acid, N,N-dimethylformamide (DMF), benzoyl peroxide (BPO), and petroleum ether were purchased from Shanghai in China. These are analytical reagents and were used as received without further purification.

Shengli (SL) waxy crude oil was selected to evaluate the efficiency of polymeric additives [35]. The physicochemical characteristics of SL crude oil are listed in Table 1.

Table 1. Physical characteristics of shengli crude oil.

Properties	Shengli Crude Oil
Density at 20 °C (g/cm^3)	0.85
Pour point (°C)	32
Wax content (wt%)	21.23
Resin content (wt%)	2.17
Asphaltene content (wt%)	0.89
IBP (°C)	60.5

2.2. Preparation of the Eight Pour Point Depressant (PPD)

Maleic anhydride can easily copolymerize with alkene monomers and react with alkyl alcohols and amines, making it possible to prepare pour point depressants by copolymerizing maleic anhydride with different monomers. We polymerized maleic anhydride with different ene monomers such as acrylic acid and acrylic ester, and added various polar and aromatic units into the polymer backbone to prepare eight copolymers with good performance (Figure 1).

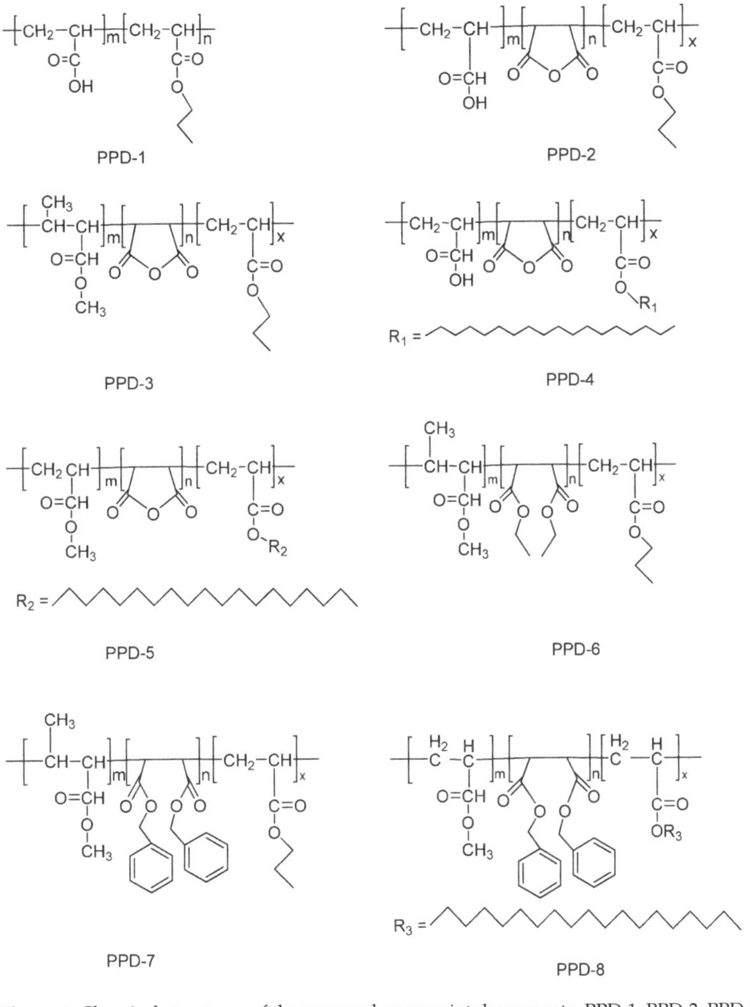

PPD-1 PPD-2 PPD-3 PPD-4 PPD-5 PPD-6 PPD-7 PPD-8

Figure 1. Chemical structures of the prepared pour-point depressants: PPD-1, PPD-2, PPD-3, PPD-4, PPD-5, PPD-6, PPD-7, and PPD-8.

2.2.1. Preparation of PPD-1

The acrylic acid and propyl acrylate (1:1, molar ratio, Table 2) were dissolved in 500 mL of solvent DMF. A mixture of 0.8 g of initiator BPO and 20 mL of petroleum ether was prepared and slowly added to the reaction vessel using a dropping funnel. The reaction was carried out under constant temperature at 80 °C for 2 h. After the polymerization reaction was complete, impurities were removed using a rotary evaporator, and the product PPD-1 was obtained after cooling to room temperature.

2.2.2. Preparation of PPD-2

The acrylic acid, the propyl acrylate, and the maleic anhydride (3:3:1, molar ratio, Table 2) were added to a three-neck flask. Solvent DMF was then poured into the flask.

During the reaction, a mixture of petroleum ether and initiator BPO was added dropwise using a dropping funnel (the amount of initiator BPO was 0.1% of the total monomer mass). The reaction was maintained at a constant temperature of 80 °C for 2 h, and the product was obtained by rotary evaporation.

Table 2. The selections of monomers.

Monomer A	Monomer B	Monomer C	Products
Acrylic acid	Propyl acrylate	None	PPD-1
Acrylic acid	Propyl acrylate	Maleic anhydride	PPD-2
Methyl methacrylate	Propyl acrylate	Maleic anhydride	PPD-3
Acrylic acid	Octadecyl acrylate	Maleic anhydride	PPD-4
Methyl acrylate	Nonadecyl acrylate	Maleic anhydride	PPD-5
Methyl methacrylate	Propyl acrylate	Diethyl succinate	PPD-6
Methyl methacrylate	Propyl acrylate	Dibenzyl maleate	PPD-7
Methyl acrylate	Nonadecyl acrylate	EsterDibenzyl	PPD-8

2.2.3. Preparation of PPD-3, PPD-4, PPD-5

The synthetic methods of PPD-3, PPD-4, and PPD-5 are similar to the preparation method of PPD-2.

2.2.4. Preparation of PPD-6

Diethyl succinate was prepared by directly reacting 1 mol of maleic anhydride with 2 mol of anhydrous ethanol in the presence of sulfuric acid (1%) as a catalyst heating at 75 °C. After the reaction was complete, the catalyst and unreacted materials were removed by washing with distilled water, and the pure ester was obtained. The methyl methacrylate, the propyl acrylate, and the pure ester prepared (3:3:1, molar ratio, Table 2) were added to a three-neck flask. Solvent DMF was then poured into the flask. During the reaction, a mixture of petroleum ether and initiator BPO was added dropwise using a dropping funnel (the amount of initiator BPO was 0.1% of the total monomer mass). The reaction was maintained at a constant temperature of 80 °C for 2 h, and the product was obtained by rotary evaporation.

2.2.5. Preparation of PPD-7

Dibenzyl maleate was prepared by directly reacting 1 mol of maleic anhydride with 2 mol of benzyl alcohol in the presence of sulfuric acid (1%) as a catalyst heating at 75 °C. After the reaction was complete, the catalyst and unreacted materials were removed by washing with distilled water, and the pure ester was obtained. The remaining steps were similar to the preparation method of PPD-6.

2.2.6. Preparation of PPD-8

Based on the synthesis of the binary copolymerized pour point depressant PPD-1, ternary copolymerized pour point depressants were obtained by inserting a third monomer, which improved the structure of PPD and had better performance. Compared with PPD-1, the structure of PPD-2 inserted maleic anhydride monomer; meanwhile, PPD-3 can be obtained by adding a methyl group in the structure of acrylic acid monomer. A higher pour point depression effect of the polymeric improver (PPD-4, PPD-5) can be obtained by exerting a long alkyl side chain on acrylic ester monomer. PPD-6 can be produced by ring-opening maleic anhydride. Aromatic rings were incorporated into ring-opening maleic anhydride groups to attain PPD-7. Both insert aromatic side chains in maleic anhydride structure and add pendant chains on acrylate monomer can fabricate PPD-8.

Generally speaking, both the copolymers prepared by us and Wu et al. [21] possessed long-chain alkyl and maleic anhydride structures. In comparison to the pour point depressants obtained by Wu et al., our PPD-1 exhibited similarities, as both were binary copolymers. Additionally, the PPD-4 we synthesized and the POM produced by Wu et al.

shared the monomeric structure of octadecyl acrylate and maleic anhydride. The presence of ring-opening maleic anhydride groups with aromatic rings and long alkyl side chains was observed in our PPD-8 and the POMB obtained by Wu et al. However, the difference lies in the fact that, apart from PPD-1, the copolymers we synthesized were ternary copolymers. We employed various ene-monomers co-polymerized with maleic anhydride, facilitating a more effective action of the pour point depressant. This was best illustrated by the outstanding performance of PPD-8 in reducing the pour point by 12 °C at a concentration of 100 ppm.

2.3. Characterization of the Products

Gel permeation chromatography (GPC) (Waters Model 1515) was used to estimate the weight-average molecular weight (Mw) of the produced PPD. Tetrahydrofuran (THF) was used as the mobile phase at a flow rate of 1 mL/min and a temperature of 30 °C. Polystyrene was used as the standard. The Fourier transform infrared spectrometry (FT-IR) analyses were measured on a Nicolet 5700 spectrophotometer (Nicolet, Thermo Fisher, Waltham, MA, USA) in the range 400–4000 cm^{-1}. ^{1}H NMR and ^{13}C NMR spectra were recorded on a 400 MHz (Bruker, Germany) instrument, and CDCl$_3$ (7.26 ppm for ^{1}H NMR, 77.16 ppm for ^{13}C NMR) was used as a reference (Supplementary Materials).

2.4. The Selection of Pour Point Depressants

Firstly, 400 mL of oil sample was taken in a conical flask and placed in a water bath at 60 °C to completely melt it, stirring evenly. Secondly, four dry and clean test tubes were taken and 10 g of crude oil was added to each tube. Three of the tubes added 0.01 g of pour point depressant as the experimental group (concentration of 100 ppm), while the remaining tube served as the control group without the PPD. They were stirred evenly and placed in a water bath at 60 °C for 2 h to allow the PPD to fully take effect. Lastly, according to the national standard GB/T3535-83 [36], the pour point test was performed on the oil samples using a KDND-801 pour point tester, the data were recorded, and the average decrease in pour point among the three test tubes was calculated as the final result. This experiment was conducted in parallel for eight sets of PPD products, identifying the PPD product with the most effective performance.

2.5. Test Concentration of Pour Point Depressants

To determine the optimal concentration of the selected PPD, the experimental procedure was as follows: Firstly, 100 mL of oil sample was added to a conical flask and placed in a water bath at 60 °C for complete melting, and it was stirred thoroughly. Then, six dry and clean test tubes were taken, and 10 g of crude oil was added to each tube. Five of the tubes were individually added with PPD at concentrations of 50 ppm, 100 ppm, 200 ppm, 300 ppm, and 500 ppm, respectively. The remaining tube served as a blank control without the addition of any PPD. After thorough stirring, the test tubes were placed in a water bath at 60 °C for 2 h to allow the PPD to fully take effect. Lastly, the pour point of the oil samples was tested using a KDND-801 pour point tester, and the data were recorded to determine the best concentration of the selected pour point depressant.

2.6. Microscopy Studies

Polarizing optical microscopy (POM) was used to examine the wax crystal structure of crude oil before and after the addition of the PPD at standard temperature, and images were captured using a computer.

3. Results and Discussion

3.1. Polymer Characterization

3.1.1. GPC of PPD

The average molecular weight (M$_w$) of the eight pour point depressants prepared was determined and the yields of these polymers are shown in Table 2. According to expert research, the molecular weight of the PPD should be within a certain range, typically between 4000 and 100,000 [30]. When the molecular weight of the PPD was below 2000, there was generally no significant effect on decreasing the pour point. Conversely, when the molecular weight exceeded 500,000, the effect was also poor for the performance activity mechanism. From Table 3, it can be observed that the average molecular weights of the PPD were all within the range of 34,958–42,339, which falls within the effective range of perfect performance. Within this range, the molecular weight was moderate, providing sufficient coverage and adhesion to effectively inhibit wax crystal growth and crystallization.

Table 3. Molecular weights and yields of pour point depressants.

Additive	M_W (g/mol)	Yield (%)
PPD-1	34,958	90
PPD-2	38,458	88
PPD-3	41,520	89
PPD-4	41,330	91
PPD-5	42,339	87
PPD-6	39,415	81
PPD-7	38,965	82
PPD-8	40,134	85

3.1.2. IR Spectroscopy Analysis of PPD

The chemical structure of the pour point depressants was determined by FT-IR spectroscopy. The FT-IR spectrum of PPD-8 is shown in Figure 2. The O-H absorption peak can be seen in 3440 cm^{-1}. The absorption peak of aromatic ring C=C was 1600 cm^{-1}. The carbonyl stretch vibration peak can be seen in 1730 cm^{-1}. The characteristic C-H of aromatic ring absorption peak at 699 cm^{-1}. It was confirmed that the acrylic ester and maleic anhydride had copolymerized and proved that we chose the right selected method to copolymerized. Additionally, the results obtained from other Fourier Transform Infrared Spectroscopy (FT-IR) analyses also demonstrated the successful execution of the synthesis process.

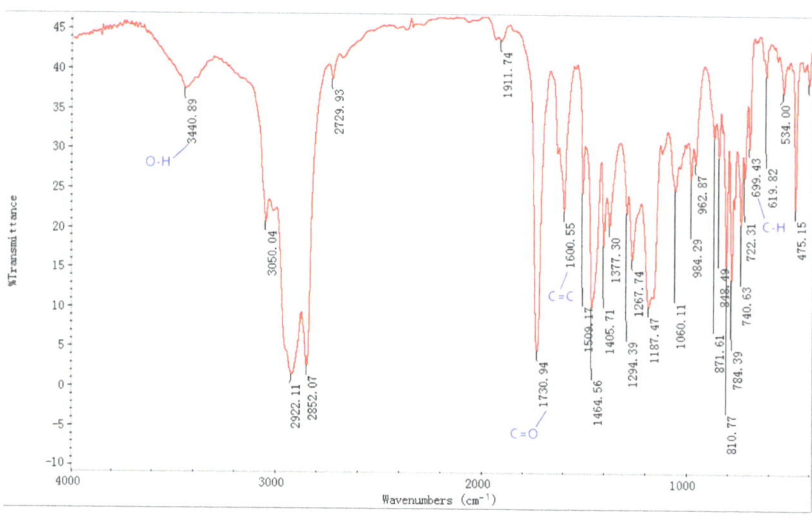

Figure 2. FT-IR spectra of PPD-8.

3.1.3. Nuclear Magnetic Resonance Spectroscopy of PPD

In order to further confirm the structure of the pour point depressants, we also obtained nuclear magnetic resonance (NMR) spectra, including ^{1}H NMR and ^{13}C NMR.

^{1}H NMR and ^{13}C NMR Analysis:

PPD-1, pale yellow oil; ^{1}H NMR: δ 3.9 (m, 1H), δ 2.2 (m, 1H), δ 1.6 (m, 2H), δ 1.2 (m, 27H), δ 0.8 (m, 2H); ^{13}C NMR: δ 32.0, 29.8, 29.7, 29.4, 25.9, 22.7, 14.2.

PPD-2, pale yellow oil; ^{1}H NMR: δ 7.4 (m, 1H), δ 7.1 (m, 1H), δ 6.4 (m, 3H), δ 6.3 (m, 2H), δ 3.3 (m, 2H), δ 3.1 (m, 5H), δ 2.9 (m, 6H), δ 2.8 (m, 2H), δ 2.1 (m, 6H), δ 1.6 (m, 2H), δ 1.3 (m, 35H), δ 0.9 (m, 3H,); ^{13}C NMR: δ 168.1, 166.0, 165.5, 136.3, 134.7, 131.3, 32.0, 29.8, 29.7, 29.4, 25.9, 22.7, 14.2.

PPD-3, pale yellow oil; ^{1}H NMR: δ 6.8 (m, 1H), δ 6.7 (m, 1H), δ 6.3 (m, 5H), δ 3.5 (m, 1H), δ 3.3 (m, 3H), δ 3.0 (m, 3H), δ 2.8 (m, 2H), δ 2.1 (m, 1H), δ 1.6 (m, 4H), δ 1.3 (m, 83H), δ 0.8 (m, 9H); ^{13}C NMR: δ 168.5, 166.0, 165.3, 136.6, 134.8, 131.0, 40.7, 37.9, 31.9, 29.7, 29.7, 29.6, 19.5, 29.4, 29.2, 28.8, 28.5, 26.8, 22.7, 14.1.

PPD-4, pale yellow oil; ^{1}H NMR: δ 6.5 (m, 2H), δ 6.3 (m, 3H), δ 6.2 (m, 3H), δ 3.4 (m, 6H), δ 3.0 (m, 1H), δ 2.2 (m, 1H), δ 1.6 (m, 6H), δ 1.3 (m, 132H), δ 0.8 (m, 12H); ^{13}C NMR: δ 136.9, 130.6, 40.7, 31.9, 29.7, 29.5, 29.4, 29.1, 28.8, 26.8, 22.7, 14.1.

PPD-5, yellow oil; ^{1}H NMR: δ 6.9 (m, 5H), δ 6.7 (m, 1H), δ 6.4 (m, 4H), δ 6.3 (m, 12H), δ 3.5 (m, 1H), δ 3.4 (m, 12H), δ 3.0 (m, 7H), δ 2.8 (m, 2H), δ 2.1 (m, 4H), δ 1.6 (m, 14H), δ 1.3 (m, 229H), δ 0.9 (m, 20H); ^{13}C NMR: δ 168.1, 166.0, 165.4, 136.5, 134.6, 134.0, 131.2, 40.7, 31.9, 29.7, 29.7, 29.6, 29.6, 29.5, 29.4, 29.2, 28.8, 26.8, 22.7, 14.1.

PPD-6, yellow oil; ^{1}H NMR: δ 7.8 (m, 3H), δ 7.4 (m, 3H), δ 7.2 (m, 1H), δ 7.0 (m, 3H), δ 4.0 (m, 1H), δ 2.6(m, 1H),δ 2.4 (m, 2H), δ 2.2 (m, 9H), δ 1.6 (m, 1H), δ 1.3 (m, 16H), δ 0.9 (m, 2H); ^{13}C NMR: δ 166.4, 137.8, 135.4, 133.5, 130.5, 130.0, 128.2, 127.9, 127.7, 127.6, 127.2, 126.9, 126.6, 126.1, 125.9, 125.7, 125.0, 124.1, 64.8, 32.0, 29.9, 29.8, 29.6, 29.5, 29.3, 28.7, 26.0, 22.8, 21.4, 19.4, 14.2.

PPD-7, yellow oil; ^{1}H NMR: δ 8.0 (m, 7H), δ 7.8 (m, 26H), δ 7.5 (m, 29H), δ 7.3 (m, 24H), δ 7.0 (m, 8H), δ 6.4 (m, 7H), δ 6.1 (m, 1H), δ 5.8 (m, 1H), δ 4.1 (m, 5H), δ 3.9 (m, 15H), δ 2.8 (m, 6H), δ 2.7(m, 16H), δ 2.5 (m, 38H), δ 2.2 (m, 44H), δ 2.0 (m, 4H), δ 1.8 (m, 4H), δ 1.6 (m, 35H), δ 1.3 (m, 305H), δ 0.8 (m, 29H); ^{13}C NMR: δ 130.5, 128.6, 128.1, 127.9, 127.7, 127.6, 127.2, 126.8, 126.5, 125.8, 125.7, 125.5, 124.9, 64.7, 31.9, 29.8, 29.7, 29.6, 29.5, 29.4, 29.3, 28.6, 25.9, 22.7, 21.7, 19.4, 14.2.

PPD-8, yellow oil; ^{1}H NMR: δ 8.0 (m, 15H), δ 8.0 (m, 15H), δ 7.8 (m, 27H), δ 7.7 (m, 54H), δ 7.3 (m, 155H), δ 6.9 (m, 42H), δ 6.4 (m, 9H), δ 6.1 (m, 9H), δ 5.8 (m, 9H), δ 4.1 (m, 22H), δ 4.0 (m, 41H), δ 2.8 (m, 10H), δ 2.6 (m, 65H), δ 2.4 (m, 103H), δ 2.2 (m, 148H), δ 1.6 (m, 102H), δ 1.2 (m, 1062H), δ 0.9 (m, 109H); ^{13}C NMR: δ 130.5, 128.9, 128.6, 128.5, 128.1, 128.9, 127.9, 127.7, 127.6, 127.2, 126.8, 126.5, 126.4, 125.8, 125.7, 125.5, 124.9, 124.1, 64.8, 32.0, 29.8, 29.7, 29.6, 29.5, 29.4, 29.4, 29.3, 28.6, 25.9, 22.7, 21.7, 20.6, 19.4, 14.2.

3.2. PPD Performance Evaluation

3.2.1. Effect of PPD Addition on Pour Point of Crude Oil

During the experiment, we investigated the effects of different types and concentrations of pour point depressants on the flowability of waxy crude oil. We placed the oil sample under 60 °C for 3 h and then allowed it to cool down steadily. We measured the effects of the eight pour point depressants on the pour point of waxy crude oil before and after addition at a concentration of 100 ppm. The results are shown in Table 4, indicating that PPD-8 had the best pour point depressant effect. Next, we measured the changes in the pour point of waxy crude oil with the addition of PPD-8 at concentrations of 50 ppm, 100 ppm, 200 ppm, 300 ppm, and 500 ppm, and the results are shown in Figure 3. It can be seen from the table that the best pour point depressant effect was achieved at a concentration of 200 ppm.

Table 4. Effect of pour point depressants at a concentration of 100 ppm on Shengli crude oil (PP = 32 °C).

Additive	Pour Point with PPD/°C	Pour-Point Depression/°C
PPD-1	26	6
PPD-2	28	4
PPD-3	27	5
PPD-4	25	7
PPD-5	24	8
PPD-6	26	6
PPD-7	23	9
PPD-8	20	12

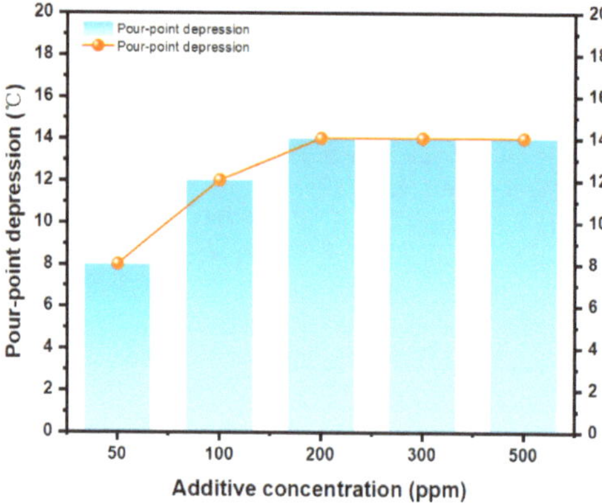

Figure 3. Pour points of shengli crude oil treated with PPD-8.

It can be distinctly observed that the pour point with pour point depressant was extremely lower than the pour point without additive PPD from Table 4, indicating that any type of pour point depressant displayed a great function on the high-pour-point crude oil. Compared with PPD-1~PPD-3, the polymer (PPD-4 and PPD-5) containing longer alkyl side chains can enhance the pour point ability by 7 °C and 8 °C, demonstrating that alkyl side chains can provide more approving adsorption sites for wax crystals. It has been shown that both PPD-7 and PPD-8 including aromatic units showed excellent performance in reducing the pour point by 9 °C and 12 °C. This can be best illustrated by the benzene ring can raise the solubility of the PPD and enhance the interactions between the PPD and paraffin in crude oil. It is not difficult to draw the conclusion that the selected PPD-8 at the concentration of 100 ppm possessed the most optimum property among the eight types of pour point depressant.

In Figure 3, it can be seen that the pour point depressant effect was poor when the concentration of the additive was low. When the concentration was increased to 200 ppm, the pour point depressant effect was better and the reduction in pour point reached 14 °C. If the concentration was further increased, the effect of the pour point depressant will not change much, which is not economically feasible and will increase the cost. Therefore, the optimal concentration for PPD-8 was 200 ppm.

Based on the above results, it can be concluded that the pour point depressant (PPD) should have appropriate alkyl chains that are hydrophobic and interact with the waxy microcrystals. The hydrophilic groups were exposed on the outer surface of the polymeric chain and acted as repellents for other waxy microcrystals, thereby inhibiting coalescence and the subsequent formation of larger waxy particles that could deposit. The long alkyl chain of the PPD played an important role in its interaction with wax crystals, as it can act as the nucleation center for wax precipitation and increase the number of small wax crystals in the crude oil, thereby preventing the formation of large network structures. Additionally, the long alkyl chain can also form eutectic with wax, altering the crystal orientation and inhibiting further wax crystal growth. Polypropylene acid ester with a long alkyl chain belongs to a comb-shaped polymer and possesses excellent shear resistance. The elongated aliphatic chains continuously extend around the resin and asphaltene, lowering the polarity of the surrounding environment and acting as a shield, thus preventing the aggregation and accumulation of resin and asphaltene fragments, resulting in a reduction in crude oil viscosity. In addition to the long alkyl chain structure, the polar functional groups in the PPD also have a significant impact on its efficiency. After the PPD is adsorbed onto the wax crystals, its polar functional groups can repel other wax crystals, preventing them from aggregating into larger clusters, thereby improving the dispersion of wax crystals and enhancing the flowability of crude oil. The polar functional groups in maleic anhydride and acrylic acid esters can effectively interact with the natural components of pour point depression of resin and asphaltene in crude oil. The PPD enters the resin and asphaltene fragments through intermolecular forces, resulting in stronger hydrogen bonding. In the meanwhile, the oxygen-containing polar functional groups also improve the PPD's resistance to repeated heating and shear stress. Moreover, the benzene ring of chains can interact with asphaltenes through a π–π stacking attractive force. The combined action of the alkane carbon chain, polar functional groups, and benzene ring prevent the mutual binding of wax crystals, thus reducing the pour point, viscosity, and yield stress of the crude oil. Therefore, the low-temperature flowability of the crude oil is significantly improved.

3.2.2. Wax Crystal Pattern of Crude Oil after Adding PPD-8

Figure 4 shows the polarized microscopy images of the crude oil without pour point depressant and with 200 ppm of pour point depressant at 20 °C. From the figure, it can be seen that the wax crystal size in the crude oil without pour point depressant was small, the wax crystals were relatively loose, and they were distributed more evenly in the oil phase. After adding PPD-8 pour point depressant, the size of the wax crystal particles in the crude oil became larger, with agglomeration forming large spherical crystal aggregates. The addition of pour point depressant changed the crystallization mode of wax in the crude oil, decreased the amount of wrapped liquid oil, weakened the flocculation ability of wax crystal aggregates, reduced the strength of the spatial grid structure, and increased the flowability of crude oil.

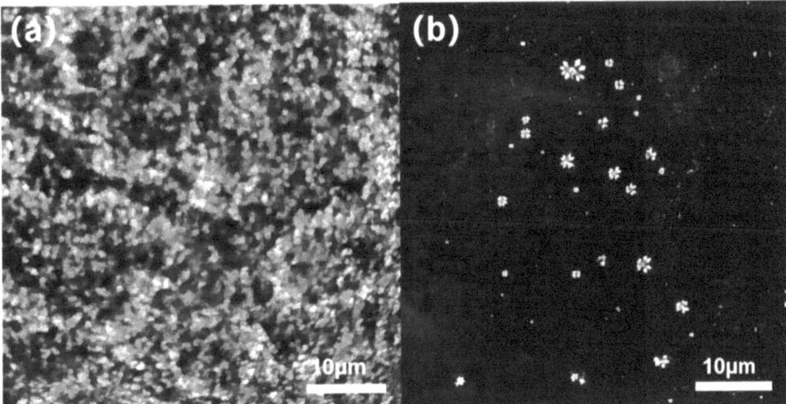

Figure 4. Polarized microscopy images of crude oil without and with PPD-8. (**a**) virgin crude oil (**b**) crude oil with 200 ppm concentration of pour point depressant at 20 °C.

3.3. Analysis of the PPD Mechanism

The long carbon chain structure in PPD easily formed eutectic with wax crystals in crude oil. The polar groups in acrylate compounds enhanced the intermolecular repulsion among the wax crystals, while the introduction of benzene rings increased the solubility of wax crystals, preventing them from agglomerating into wax deposits and enhancing the anti-deposition capability. The addition of PPD changed the crystallization pattern of the crude oil, causing the wax crystals to tend to aggregate into clusters, lowering the surface energy and reducing the spatial structural strength. This improvement at a macroscopic level manifested as an enhancement in the low-temperature flowability of the crude oil.

4. Conclusions

This study demonstrated that modified maleic anhydride co-polymers can be available pour point depressants for waxy crude oil. In this study, eight different types of pour point depressants were synthesized by copolymerizing maleic anhydride with different monomers and characterized using techniques such as FT-IR and NMR. The results showed that the addition of pour point depressants successfully reduced the pour point of the waxy crude oil, with an optimal concentration of 200 ppm for some of the tested compounds. The addition of pour point depressants changed the crystallization mode of wax in the crude oil, resulting in larger and more densely packed wax crystal aggregates, which improved the flowability of the crude oil. The study also found that the performance of the pour point depressants can be optimized by changing the hydrophobic parts and hydrophilic part of the polymer structure. Therefore, modified maleic anhydride co-polymers have the potential to be used as effective pour point depressants for waxy crude oil.

Supplementary Materials: The following supporting information can be downloaded at: https://www.mdpi.com/article/10.3390/polym15193898/s1. Ref. [37] is cited in the supplementary materials.

Author Contributions: Investigation, Q.L. and H.C.; Methodology, J.Y. and R.L.; Project administration, Y.G. and C.J.; Supervision, B.Y. and D.Y.; Writing—original draft, W.Z. All authors have read and agreed to the published version of the manuscript.

Funding: We are grateful to National Innovation and Entrepreneurship Training Program for College Students (CXCY2023105, S202210447044, 202210447015) for financial support of this research.

Data Availability Statement: The data used to support the findings of this study are available from the corresponding author upon request.

Conflicts of Interest: The authors declare no conflict of interest.

References

1. Lv, G.; Li, Q.; Wang, S. Key techniques of reservoir engineering and injection–production process for CO_2 flooding in China's SINOPEC Shengli Oilfield. *J. CO_2 Util.* **2015**, *11*, 31–40. [CrossRef]
2. Weidong, W.; Junzhang, L.; Xueli, G. MEOR field test at block Luo801 of Shengli oil field in China. *Pet. Sci. Technol.* **2014**, *32*, 673–679. [CrossRef]
3. Gao, C.; Shi, J.; Zhao, F. Successful polymer flooding and surfactant-polymer flooding projects at Shengli Oilfield from 1992 to 2012. *J. Pet. Explor. Prod. Technol.* **2014**, *4*, 1–8. [CrossRef]
4. Irfan, M.; Khan, J.A.; Al-Kayiem, H.H. Characteristics of Malaysian Crude Oils and Measurement of ASP Flooded Water in Oil Emulsion Stability and Viscosity in Primary Separator. *Water* **2023**, *15*, 1290. [CrossRef]
5. Ashoori, S.; Sharifi, M.; Masoumi, M. The relationship between SARA fractions and crude oil stability. *Egypt. J. Pet.* **2017**, *26*, 209–213. [CrossRef]
6. Yasin, G.; Bhanger, M.I.; Ansari, T.M. Quality and chemistry of crude oils. *J. Pet. Technol. Altern. Fuels* **2013**, *4*, 53–63.
7. Xu, J.; Xing, S.; Qian, H. Effect of polar/nonpolar groups in comb-type copolymers on cold flowability and paraffin crystallization of waxy oils. *Fuel* **2013**, *103*, 600–605. [CrossRef]
8. Kurniawan, M.; Norrman, J.; Paso, K. Pour point depressant efficacy as a function of paraffin chain-length. *J. Pet. Sci. Eng.* **2022**, *212*, 110250. [CrossRef]

9.   Ma, R.; Zhu, J.; Wu, B. Distribution and qualitative and quantitative analyses of chlorides in distillates of Shengli crude oil. *Energy Fuels* **2017**, *31*, 374–378. [CrossRef]

10.   Li, H.; Liu, Y.; Luo, X. A novel nonlinear multivariable Verhulst grey prediction model: A case study of oil consumption forecasting in China. *Energy Rep.* **2022**, *8*, 3424–3436. [CrossRef]

11.   Yi, S.; Zhang, J. Relationship between waxy crude oil composition and change in the morphology and structure of wax crystals induced by pour-point-depressant beneficiation. *Energy Fuels* **2011**, *25*, 1686–1696. [CrossRef]

12.   He, C.; Ding, Y.; Chen, J. Influence of the nano-hybrid pour point depressant on flow properties of waxy crude oil. *Fuel* **2016**, *167*, 40–48. [CrossRef]

13.   Mohanan, A.; Bouzidi, L.; Narine, S.S. Harnessing the synergies between lipid-based crystallization modifiers and a polymer pour point depressant to improve pour point of biodiesel. *Energy* **2017**, *120*, 895–906. [CrossRef]

14.   Adams, J.J.; Tort, F.; Loveridge, J. Physicochemical Approach to Pour Point Depressant Treatment of Waxy Crudes. *Energy Fuels* **2023**, *37*, 6432–6449. [CrossRef]

15.   Kamal, R.S.; Shaban, M.M.; Raju, G. High-Density Polyethylene Waste (HDPE)-Waste-Modified Lube Oil Nanocomposites as Pour Point Depressants. *ACS Omega* **2021**, *6*, 31926–31934. [CrossRef] [PubMed]

16.   Wang, C.; Zhang, M.; Wang, W. Experimental study of the effects of a nanocomposite pour point depressant on wax deposition. *Energy Fuels* **2020**, *34*, 12239–12246. [CrossRef]

17.   Jia, X.; Fu, M.; Xing, X. Submicron carbon-based hybrid nano-pour-point depressant with outstanding pour point depressant and excellent viscosity depressant. *Arab. J. Chem.* **2022**, *15*, 104157. [CrossRef]

18.   Xue, Y.; Chen, F.; Sun, B. Effect of nanocomposite as pour point depressant on the cold flow properties and crystallization behavior of diesel fuel. *Chin. Chem. Lett.* **2022**, *33*, 2677–2680. [CrossRef]

19.   Li, H.; Chen, C.; Huang, Q. Effect of pour point depressants on the impedance spectroscopy of waxy crude oil. *Energy Fuels* **2020**, *35*, 433–443. [CrossRef]

20.   Li, X.; Yuan, M.; Xue, Y. Tetradecyl methacrylate-N-methylolacrylamide Copolymer: A low concentration and high-efficiency pour point depressant for diesel. *Colloids Surf. A Physicochem. Eng. Asp.* **2022**, *642*, 128672. [CrossRef]

21.   Yang, F.; Zhao, Y.; Sjöblom, J. Polymeric wax inhibitors and pour point depressants for waxy crude oils: A critical review. *J. Dispers. Sci. Technol.* **2015**, *36*, 213–225. [CrossRef]

22.   Li, N.; Mao, G.L.; Shi, X.Z. Advances in the research of polymeric pour point depressant for waxy crude oil. *J. Dispers. Sci. Technol.* **2018**, *39*, 1165–1171. [CrossRef]

23.   Hao, L.Z.; Al-Salim, H.S.; Ridzuan, N. A Review of the Mechanism and Role of Wax Inhibitors in the Wax Deposition and Precipitation. *Pertanika J. Sci. Technol.* **2019**, *27*, 499–526.

24.   Liu, T.; Fang, L.; Liu, X. Preparation of a kind of reactive pour point depressant and its action mechanism. *Fuel* **2015**, *143*, 448–454. [CrossRef]

25.   Huang, H.; Wang, W.; Peng, Z. The influence of nanocomposite pour point depressant on the crystallization of waxy oil. *Fuel* **2018**, *221*, 257–268. [CrossRef]

26.   Cao, J.; Liu, L.; Liu, C. Phase transition mechanisms of paraffin in waxy crude oil in the absence and presence of pour point depressant. *J. Mol. Liq.* **2022**, *345*, 116989. [CrossRef]

27.   El Mehbad, N. Efficiency of N-Decyl-N-benzyl-N-methylglycine and N-Dodecyl-N-benzyl-N-methylglycine surfactants for flow improvers and pour point depressants. *J. Mol. Liq.* **2017**, *229*, 609–613. [CrossRef]

28.   Yao, B.; Li, C.; Yang, F. Organically modified nano-clay facilitates pour point depressing activity of polyoctadecylacrylate. *Fuel* **2016**, *166*, 96–105. [CrossRef]

29.   Oliveira, L.M.S.L.; Nunes, R.C.P.; Melo, I.C. Evaluation of the correlation between wax type and structure/behavior of the pour point depressant. *Fuel Process. Technol.* **2016**, *149*, 268–274. [CrossRef]

30.   Xie, M.; Chen, F.; Liu, J. Synthesis and evaluation of benzyl methacrylate-methacrylate copolymers as pour point depressant in diesel fuel. *Fuel* **2019**, *255*, 115880. [CrossRef]

31.   Steckel, L.; Nunes, R.C.P.; Rocha, P.C.S. Pour point depressant: Identification of critical wax content and model system to estimate performance in crude oil. *Fuel* **2022**, *307*, 121853. [CrossRef]

32.   Xue, Y.; Zhao, Z.; Xu, G. Effect of poly-alpha-olefin pour point depressant on cold flow properties of waste cooking oil biodiesel blends. *Fuel* **2016**, *184*, 110–117. [CrossRef]

33.   Wu, Y.; Ni, G.; Yang, F. Modified maleic anhydride co-polymers as pour-point depressants and their effects on waxy crude oil rheology. *Energy Fuels* **2012**, *26*, 995–1001. [CrossRef]

34.   Fang, L.; Zhang, X.; Ma, J. Investigation into a pour point depressant for Shengli crude oil. *Ind. Eng. Chem. Res.* **2012**, *51*, 11605–11612. [CrossRef]

35.   Xu, G.; Xue, Y.; Zhao, Z. Influence of poly (methacrylate-co-maleic anhydride) pour point depressant with various pendants on low-temperature flowability of diesel fuel. *Fuel* **2018**, *216*, 898–907. [CrossRef]

36. *GB/T3535-83*; Determination of Oil Pour Point. China Standards Press: Beijing, China, 1983.
37. Pucko, I.; Racar, M.; Faraguna, F. Synthesis, characterization, and performance of alkyl methacrylates and tert-butylaminoethyl methacrylate tetra polymers as pour point depressants for diesel Influence of polymer composition and molecular weight. *Fuel* **2022**, *324*, 124821. [CrossRef]

MDPI AG
Grosspeteranlage 5
4052 Basel
Switzerland
Tel.: +41 61 683 77 34

Polymers Editorial Office
E-mail: polymers@mdpi.com
www.mdpi.com/journal/polymers